NAVY
STRATEGIC CULTURE

NAVY STRATEGIC CULTURE

Why the Navy Thinks Differently

Roger W. Barnett

NAVAL INSTITUTE PRESS
Annapolis, Maryland

Naval Institute Press
291 Wood Road
Annapolis, MD 21402

Library of Congress Cataloging-in-Publication Data

Barnett, Roger W.
Navy strategic culture : why the Navy thinks differently/Roger Barnett.
p. cm.
Includes bibliographical references and index.
ISBN 978-1-59114-024-5 (alk. paper)
1. United States. Navy. 2. Strategic culture–United States. 3. United States. Navy–Officers. 4. Sociology, Military–United States. 5. Naval art and science–United States. 6. Naval strategy. 7. National security–United States. I. Title.
VA55.B33 2009
359'.030973–dc22

2009027768

Printed in the United States of America on acid-free paper ∞

14 13 12 11 10 09 9 8 7 6 5 4 3 2
First printing

To those who go down to the sea in fighting ships . . .

Contents

Illustrations

Figures

Tables

Preface

Upon reporting to my first ship, a World War II–vintage destroyer, I was instructed by the ship's operations officer—a crusty mustang (up-through-the-ranks) lieutenant—as follows: "Son, if you want to be successful in this man's Navy, you need to do only three things: Keep your sides clean, shoot straight, and stay off the ^#%*§ radio!" I know now, but could only sense dimly then, that his tagline referred idiomatically to order and discipline, fighting spirit, and self-reliance. These are three of the core cultural concepts of the U.S. Navy.

My second ship was a postwar destroyer named *Forrest Sherman* after the admiral who was the chief of Naval Operations during the Korean War. The motto of *Forrest Sherman* was: "A home, a feeder, and a squadron leader." Ships are truly home to sailors—an oceangoing home charged, according to Title 10 of the U.S. Code with the responsibility of being organized, trained, and equipped "primarily for prompt and sustained combat incident to operations at sea."

What a different world—that of seagoing warriors! The engineering department of the aircraft carrier USS *Carl Vinson* (CVN-70) twenty years ago sent a message to the USS *Samuel B. Roberts* (FFG-58) after that ship had succeeded in coping with the damage it suffered from striking an Iranian sea mine in the Persian Gulf. The note expressed "feelings of solidarity" with the crew of the stricken frigate, and then asserted: "We think of the Navy Hymn and Naval History, and we know that only a few of our countrymen fully understand what we do and why we do it."[1] It's true: Americans are notoriously disinterested in history, and even less so in maritime affairs. As one observer wrote: "The endeavors of naval officers . . . tend not to resonate with ordinary citizens in the same way as the exploits of soldiers and Marines."[2] The intention of this book is to offer some insights into Navy Strategic Culture in order to improve that understanding.

In retrospect, my background was eminently suitable for naval service. Born into a middle-class family, raised in austere conditions, I attended an all-male public high school in a big city. The first person on either side of my family to graduate from high school, I then entered the ranks of what has become another very rare species: a Navy ROTC graduate from an Ivy League university. Virtually all of my close friends while I was growing up were male. This I consider to have been a great benefit in my formative years, for it permitted me to focus and dedicate myself to sports and to my studies, I am convinced that I would not have performed nearly as well in high school or college had I been exposed during those years to the diurnal tensions and stresses of sexual attraction and competition. Accustomed to male bonding and to interacting only sporadically with females my own age, I adapted easily to Navy life. Yet, I was married—to my college sweetheart—for my entire Navy career.

All of my sea duty was served in cruisers and destroyers. Almost all of my shore duty, and fully 30 percent of my service, was on assignment in the Pentagon. It was there that I discovered how different Navy Strategic Culture was from the cultures of the other services. As Don Snider has written: "It should be obvious to any observer, not to mention participant, that the army, navy, air force, and marine corps display sharply divergent cultures. Derived over time from their assigned domain of war on land, sea, and in the air, these individual services have developed very different ideals and concepts that in turn strongly influenced their institutional cultures and behavior, particularly their strategic approach to war that establishes their claim on the nation's assets."[3] This book is my attempt to share what I know about Navy Strategic Culture, its value and applicability to the contemporary world, and how it labors under the assaults of a post-modern, post–Cold War international environment.

Navy Strategic Culture is primarily for, and about, "those who go down to the sea in ships." It is also for those who wear the Navy uniform whose primary duties do not involve shipboard life, as well as for those who do not, have not, and will never go to sea in ships. The point of view is that of a career Navy officer. It is the officer corps that thinks about how to conduct warfare: about how the Navy can make a strategic difference. It is the officer corps that adopts and adapts the concepts to suit the salt water environment in which they are applied, for "a military can only be considered professional so long as the vast majority of its officers are loyal to its ethos."[4]

The Navy officer corps is trifurcated into those who fly aircraft, and those who sail ships and submarines. Of course, the aviators and submariners are rather new on the scene; ship drivers have been around since the fifth century BCE. Nonetheless, one of the premises of *Navy Strategic Culture* is that when it comes to strategic culture, the three groups are in close, and even comfortable, agreement.

As I drafted the manuscript for this book I consulted with officers from the separate communities—air, surface, and sub-surface—to confirm my understanding of the scope of their concurrence. In this regard, my gratitude for their patience and wise counsel goes unstintingly to Rear Adms. Jerry Holland, Dennis McCoy, Jim Stark, and Joe Strasser, and Capts. John Bonds, Jim Brick, Mike Caldwell, Don Gentry, Max Poirrier, and Mike Tollefson. Errors in the book remain in spite of their counsel, clearly not because of it.

I would be negligent if I did not mention the devotion and unflagging support of my first (and only) wife, Sandy. I am indebted to her in so very many ways.

Finally, all need to be reminded that "The military are the courage element in our community. . . . They do more than 'serve'; they are our guardians. . . . Courage is in our nature if only we look for it, but the next time it may not be ready to hand if we continue trying to suppress it. . . . Courage in Greek is also the word for manliness."[5]

CHAPTER 1

Introduction

In the late winter of 2006–2007 Adm. William Fallon was nominated to become commander of the U.S. Central Command. Admiral Fallon at that time commanded the U.S. Pacific Command—the largest of the U.S. unified command areas—one traditionally (since 1947) commanded by a Navy officer, while the Central Command had never been. Previously, in October 2006, Adm. James Stavridis assumed command of the U.S. Southern Command. He was the first Navy officer to hold that post, traditionally filled by Army officers. Then, in June 2007, Adm. Michael Mullen was nominated to replace the chairman of the Joint Chiefs of Staff, Marine Corps Gen. Peter Pace. It was said that the Bush administration opted to make a change in order to avoid a hostile Senate reconfirmation of General Pace. In July 2007 a Navy officer, Adm. Eric Olson, was named to head the Joint Special Operations Command, the first Navy man to lead this joint command, also customarily headed by Army officers. Why should Navy officers be selected for these key posts at this crucial time?

It seems that the ability of Navy officers to think strategically, to rise above the minutiae of the tactical battlefield, and to discern "where the big picture fits in" rendered them uniquely valuable as combat commanders in that particular global security environment. Of interest, of the eighteen U.S. presidents in the twentieth century, six (Kennedy, Johnson, Nixon, Ford, Carter, and G. H. W. Bush) served on active duty or as a reservist with the Navy. Two (the Roosevelts) were assistant secretaries of the Navy, and six have had aircraft carriers named after them (Theodore and Franklin D. Roosevelt, Truman, Kennedy, Reagan, and G. H. W. Bush). Whereas an Army officer as the commander of the Central Command in early 2007 would likely have become narrowly focused on the ground operations in Iraq and Afghanistan, a Navy officer was perceived as one who could lift his sights above those conflicts in order to assess and deal with emerging strategic situations in, for example, Iran, Saudi Arabia, Egypt, Syria, and Kazakhstan—all of which lie within the Central Command's area of responsibility. (See Figure 1). As one observer noted years ago: "Practical common sense and the ability to improvise when short on doctrine or material are equally essential for survival at sea. . . . Relatively unaffected by shifting political winds ashore, the sailor fashions a hardened sense of duty and loyalty to his ship and his profession . . . broadly speaking, sea power works slowly and subtly, whereas generals, politicians and the people at large are impatient for direct, immediately apparent results, as with armies on the march."[1]

Indeed, the terms "strategic," "operational," and "tactical" as levels of warfare are differently understood and assimilated in the three U.S. armed services. For Army strategists, levels of warfare are established by the *geographic scope of conflict*. Hence, the terms can be explained essentially in terms of distance:

strategic = *inter*continental;
operational, or theater = *intra*continental; and
tactical = battlefield.

For Air Force strategists, levels of warfare express the *intensity of conflict*:

strategic = nuclear;
operational and tactical are progressively lesser levels of non-nuclear conflict, differentiated by scale.

For Navy strategists, levels of warfare tend to be defined by *objectives*. Combat that involves war-winning objectives are strategic; those that are concerned with campaigns, operational; and those that seek to win battles are tactical. Hence:

strategic = "the level at which wars are decided";
operational, "the level at which campaigns are decided"; and
tactical, "the level at which battles are decided."

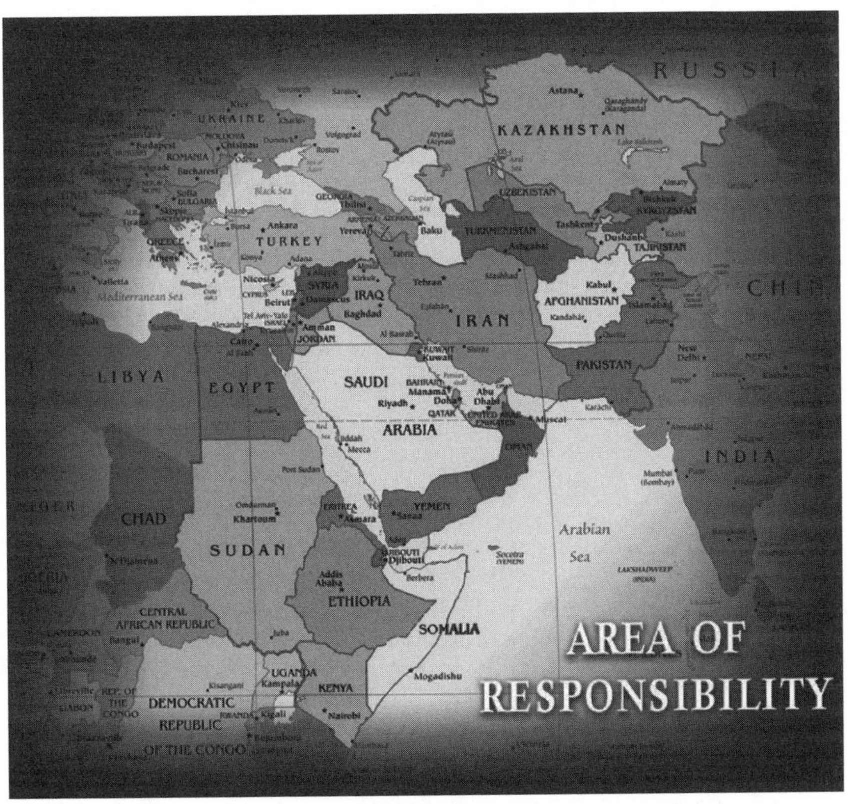

Figure 1. Central Command (CENTCOM) geographic area of responsibility.

"One could say that Halsey was the man to win a battle for you, Spruance was the man to win a campaign, but Nimitz was the man to win a war."[2]

A "strategic weapon" in Army parlance is one that traverses continents; for the Air Force it's a nuclear missile or bomber; but the Navy holds that it's one that can, by successfully attacking a vital objective, resolve wars. Thus, for a Navy strategist a knife in the hand of an assassin in the bedchamber of a head of state can constitute a strategic weapon because, on its own, it can determine the course or outcome of a war. Heads of state are *strategic targets*, and killing them might constitute a strategic objective.

In the strategic context of the twenty-first century, geography interposes few restrictions on addressing strategic targets directly and there is no hard requirement for a time-sequencing of actions to achieve success. This indicates that the levels of warfare have lost much of their particularity; they are becoming both less distinct and significantly foreshortened. More and more, it is becoming possible to invoke decisive strategic effect with tactical level actions. Because of the speed of modern combat operations, because of the increased striking power that has devolved from the sole province of the sovereign state down to the level of the individual fighter, because of the increased difficulty in providing and maintaining a secure sanctuary for any potential target, because of the concurrent improved capacity to find and identify critical targets, and because of the rapid dissemination of information to all echelons, the ability to create *strategic* effects by *tactical* and *operational* level actions has increased dramatically in recent decades.

This has resulted in a flattening, a compression, and a blurring of the lines between the levels of war. So, for one who thinks like a Navy officer, warfare has become more and more holistic. It is far more complex than distance or levels of intensity might imply.

In part, these differences in conceptualization about the levels of war stem from the variability in the way the different services have been employed by U.S. decision makers over the years. Clearly, long-range nuclear weapons are so potentially destructive and dangerous that they have been essentially removed from the notion of combat at arms. Their use was too horrible, and the consequences too severe, to contemplate except in the most extreme circumstances; and, fortunately, the United States generally commanded an advantage against possible opponents in its ability to produce and field strategic weapons of all types, undergirding the notion that nuclear deterrence was stable.

The political leadership of the United States has demonstrated repeatedly that it would suffer battlefield casualties among its armed forces rather than even to threaten nuclear employment. Political uses for nuclear weapons have scant credibility as a consequence. Moreover, U.S. decision makers do not think of the U.S. nuclear arsenal as offering a war-winning capability. U.S. governments have not required that their nuclear forces be able to prevail in a warfighting environment. In the language of strategic theorists such was termed "counterforce dominance," and was said to be "destabilizing." Thus, a deterrent standoff, based on "parity," was deemed to be sufficient.

Taken in broad strategic perspective, this is historically fascinating. That a state, challenged to its very existential roots, would stay its hand in using weapons of which it has many, that would have a high probability of ending conflict quickly and decisively, and that such weapons would have been withheld even while its fighting forces are offering the ultimate sacrifice on the battlefield will be the stuff of retrospective examination by future generations—if any records survive to study. This occurs even while there is greatly elevated concern and trepidation that nuclear weapons, and other powerful weapons and devices that would cause significant death and destruction, in the hands of a different set of actors could signal "the end of the world as we know it" and a headlong reversion to barbarism.

Likewise, the employment of U.S. ground forces abroad in warfighting mode has been undertaken spasmodically, but only with great reservation; and, since World War II, with decidedly mixed results. Successful operations include the 1983 operation in Grenada during the first Reagan administration, the late 1989–early 1990 Operation Just Cause against Panama, and the 1991 Gulf War against Saddam Hussein's Iraq. The latter operations both occurred in the administration of President George H. W. Bush. The president, however, was tarnished by the aftermath in which Shiites in southern Iraq and Kurds in northern Iraq were slaughtered by the Iraqi leader, resulting in a twelve-year imposition of no-fly zones enforced by U.S. airpower (Operations Northern Watch and Southern Watch).

In contrast, the Navy has been employed abroad almost continually since World War II: in preparations for war against the Soviet Union, in limited combat, for humanitarian relief and noncombatant evacuation, to underwrite international free use of the open oceans, and for political leverage in peacetime. U.S. decision makers consider the Navy the first line of U.S. active defense. Because of its flexibility, its sustainability, its ability to deploy autonomously, and to retract without political liability, the Navy has been the U.S. force of choice for peacekeeping and crisis response since the end of World War II.

A major downside of this approach to national security is that it creates a peacetime atmosphere of detachment and comfortable sanctuary. In the early twentieth century the French ambassador to the United States noted that America was blessed by having militarily weak neighbors to the north and south, "on the east, fish, and on the west, fish."[3] Quite a large role has been identified for this geopolitical situation: "Maritime insularity, then, has been a key ingredient in intellectual ferment, the growth of applied technology, and the fostering of democracy. Isolated and thereby well-defended men have tended to be free men, free to think and to apply their ideas to machinery and to government."[4] Geopolitically, almost all states of the world look first and foremost to their territorial integrity—their security from attack from their neighbors. Except for a few minor incursions during World War II, the mainland United States has suffered a major attack from foreign sources only once since the War of 1812, the one that occurred on September 11, 2001.

When the United States is engaged in armed conflict the Navy is required by U.S. strategy to prevail in its environment. While the nuclear forces can tolerate a standoff, and while the West might pray that its conventional ground and air forces can perform well enough to thwart conventional aggression abroad or discourage the export of terrorism without undertaking the burdens of empire, it is considered imperative that the Navy control the seas and use them purposefully for the projection of power ashore. U.S. naval forces can be counted upon to support and operate with allied naval forces all around the globe in the event of armed conflict. And, absent a successful naval campaign, other U.S. forces would be unable to pursue their own missions. If the use of nuclear weapons cannot be tolerated either for warfighting or political suasion, the use of naval forces makes sense only in those contexts. Naval forces in offensive roles are not solely for engaging in battle with or deterring other naval forces. Their function is to influence what happens "at the seat of purpose"—on the land.[5]

The key contribution of the Navy to U.S. national security is the conduct of operations *at* and *from* the sea, today and in the future. Neither analogues nor substitutes exist. Such operations, and the professionalism needed to perform them well, are truly unique. Many nations have maintained large armies, some have developed capable air forces, but the number of states that have supported large, proficient navies—national navies having appeared only at the end of the fifteenth century—can be counted on one's fingers. Such special capabilities and the mind-set they foster offer some insight into how and why Navy officers think differently from others.

This unique global potential possessed by the U.S. Navy exposes the shallowness of the objections raised within the U.S. security community against the constitutional requirement "to provide and maintain a Navy." For example, officers of the fledgling Air Force in the late 1940s were saying: "Why should we have a Navy at all? . . . There are no enemies for it to fight except apparently the Army Air Force."[6] And, "To maintain a five-ocean Navy to fight a no-ocean opponent . . . is a foolish waste of time, men, and resources."[7]

Yet, a distinctively naval approach to thinking about the application of military force in general, and naval force in particular, is not a recent phenomenon. It was apparent from the very beginning, during the initial interactions of the new United States of America with the world outside the North American continent. The naval action against the Barbary pirates was the first expeditionary military operation engaged in by U.S. forces outside the Western Hemisphere. In a manner that today seems almost surreal, it involved the application of force against Muslim extremists. Given that the operations were grounded in political, economic, and religious issues, it was controversial; and it was considered by critics to be somewhat remote from U.S. core security.

An "approach to thinking about the application of military force" is not an unconventional way to describe a "strategic culture." This book analyzes the unique strategic culture of the U.S. Navy, its viewpoints and expectations, and how it arose and changed over time. It examines the strategic approach of the U.S. Navy, and in doing so may reveal why today it is Navy officers who have grasped the top military positions in the U.S. national security structure.

Chapter 2 focuses on military culture, and then exclusively on Navy culture. It describes how culture is acquired, shaped, nurtured, and transmitted. The particular characteristics of Navy culture, its manifestations, and impact are detailed.

The unique maritime context, the key to understanding Navy Strategic Culture, is examined in Chapter 3. Here the arcane matters faced by mariners confronted moment to moment by a hostile environment are illuminated. The physical environment is examined first, because it has the greatest impact on those who go down to the sea in ships. Then, in order, the political, legal, and economic realms pertinent to sea power are discussed. The open oceans are truly unknown to most people, but their effect on those who sail them as a profession—physically, psychologically, and culturally—is profound.

Building on the description of the strange and unique environment in which naval operations are conducted, Chapter 4 considers the range of options available to practitioners of "naval art." The term "naval art" is used to suggest that, as in the practice of medicine, science underwrites the enterprise, but the application draws as much—or more—upon art as upon science. Military strategy focuses on the application of force to attain military, and ultimately political, objectives. Chapter 4 examines the options that have been available, and those that will be available in the future for the application of naval power.

The central theme of U.S. naval operations since the inception of the U.S. Navy and Marine Corps can be summarized in a word: "expeditionary." Chapter 5 focuses on the what and the how of "expeditionary" as the core concept for the naval services. The key parts of "expeditionary"—forward, mobile, offensive, self-reliant, and adaptable—are appraised in detail.

Chapter 6 looks intensively at the relationship between Navy Strategic Culture and technology. Clearly, the relationship is close, and it is a powerful factor driving Navy Strategic Culture. It explains why the concepts of "systems of systems" and "networkcentric operations" had to have been originated by Navy officers.

As the Cold War entered its later stages, even though the participants were unaware of it, the Navy stepped forward with a new "Maritime Strategy." It incorporated all the elements, long espoused but not fully realized, of Navy Strategic Culture. Chapter 7 recounts how the Maritime Strategy of the 1980s pulled all the loose ends together into a coherent, truly strategic, approach to conflict with the Soviet Union.

Before summarizing and concluding in Chapter 9, a retrospective is undertaken in Chapter 8 to address countercultural issues—ones with the potential to thwart the full realization of the culture—and other matters that might have been inadequately covered in, or omitted from, earlier chapters.

The continuity in thought, and the stability of the Navy Strategic Culture over the decades, from the shores of Tripoli to the banks of the Tigris, have been remarkable. Remarkable, yes, and understandable insofar as they were so heavily influenced by the environment—the sea—and crafted by men with salt water in their veins.

CHAPTER 2

Strategic Culture

Perhaps not oceans, but certainly lakes of ink have been devoted to describing and explaining different aspects of culture. It is generally agreed that strategic culture—as used here, and differentiated from organizational, social, political, ethnic, or the myriad of other subdivisions of the subject—consists of shared beliefs, values, and habits among persons in a military or paramilitary organization with regard to the use of military force. It is "strategic" because the function of "strategies" is to detail the "ways" particular military "means" are to be employed to achieve certain "ends."

The culture is shared among group members, and it is formed and shaped by the consideration of history, an embrace of common understandings, and the partaking of mutual experiences. The strategic culture is reflected in the attitudes and actions of those who share it. Members of a strategic culture are essentially, although not totally or thoroughly and perhaps unconsciously, like-minded on the core issues and values that unite their culture. Practitioners are *professionals*, meaning that they can be expected to behave consistently and reliably in accordance with their culture.[1] This is the sense in which Navy strategic culture is examined here.

Within the U.S. Navy "line," distinguished from the "staff" corps: the "line" encompasses those officers "eligible for command at sea," whereas the "staff" is comprised of supply, civil engineering, judge advocate general, medical, medical service, dental, and chaplain corps. Naval officers also serve in special operations roles, "SEALs," who maintain an entirely separate and very different strategic culture. Aside from the SEALs, three professional subcultures co-exist: surface ship, submarine, and aviation. After commissioning, Navy line officers follow one of those three career paths. The three subcultures have their discrete organizational and social cultures, yet their strategic culture—conditioned as it is by the same factors pertaining to the application of military force—are identical in all important respects. To repeat: those factors are an understanding and appreciation of relevant history, an embrace of common understandings, and the sharing of mutual experiences.

An appreciation for naval history is part of every midshipman's and every Navy officer candidate's preparation for commissioning, and naval history is subsequently interspersed in their career training and education. The U.S. Navy has a long history of employment as a first line of defense.[2] It has been deployed around the world from the earliest times of the republic, fought in every war in which U.S. military forces have been engaged, and enjoys a most successful heritage. The central themes of U.S. Navy strategic culture have strong historical roots, a brief historical tracing of which will be undertaken in Chapter 7.

Among the common understandings that inform the strategic culture is the place of the military in U.S. society. Within a democracy like that of the United States, one that has imposed firm and unbroken civilian control over its military, there must be embedded within the military culture a code by which it becomes self-restraining. Nevertheless, all non-military governments—even democracies—harbor an innate fear of standing armies, and this is one of the reasons they are shunned in the U.S. Constitution. For example, in 1958 during U.S. Senate hearings on defense reorganization, senators "raised the specter of the man on horseback, a military leader who might threaten civil liberties and the viability of the Republic. Such critics held that a commander responsible for the homeland and authority over CONUS-based (CONUS = Continental United States) forces or a strong chairman [of the Joint Chiefs of Staff] with a general staff and operational authority could represent the threat to the government that the founding fathers sought to avoid through militias and a constitutional proscription against large standing armies."[3]

Fear of the military by the citizenry is, of course, historically well founded. History is rife with examples of military organizations ascending to power and ruling independent states. The Japanese film *The Seven Samurai*, re-made in the United States as *The Magnificent Seven*, however, provides an avenue to a deeper understanding of the issue.[4]

Briefly, the scenario is this: poor peasants are being suppressed and exploited by a ruthless band of cutthroats. The Seven come to town, rout the villains, and then—to the great astonishment of the oppressed—accept only modest recompense for their efforts and, most significantly, they depart. The liberated souls are amazed, for they expected—having been conditioned by history and experience—that the Seven would become their new masters. But unlike other conquerors, the Seven were motivated neither by greed nor a lust for power, but to do what was right. That constituted the code under which they acted.

Today's U.S. military is bound by a similar code. When military officers receive their commissions, they are obligated to take an oath, which reads: "I, _______, having been appointed an officer in the Army [Navy, Marine Corps, Air Force, or Coast Guard] of the United States, as indicated above in the grade of ________ do solemnly swear (or affirm) that I will support and defend the Constitution of the United States against all enemies, foreign or domestic, that I will bear true faith and allegiance to the same; that I take this obligation freely, without any mental reservations or purpose of evasion; and that I will well and faithfully discharge the duties of the office upon which I am about to enter; So help me God."[5] Furthermore, all service members are bound by a code of conduct that sets forth the behavior and obligations expected of them during combat or captivity. Article VI of the code states: "I will never forget that I am an American fighting man, responsible for my actions, and dedicated to the principles which made my country free."[6] These are but two of the ways a democratic society seeks to check the inherent power of the military, which it is unwilling to trust without serious reservations.

There are other ways the military is hobbled so that it can be trusted not to threaten the democratic society it is sworn to defend. One is to hold it to a higher moral standard than other communities in the society. This accounts for a separate military judicial system, within which service members are subject to the Uniform Code of Military Justice. An entire judicial system, distinct from the federal judicial system, has been established to adjudicate transgressions against this code. Of interest, the code casts a wide net in an effort to bound and underwrite an atmosphere of order—the establishment and maintenance of discipline— listing as transgressions against the code such actions as malingering; sodomy; unauthorized absence; contempt, disrespect, or insubordinate conduct toward a superior; drunk on duty; provoking speech or gestures; making, drawing, or uttering check, draft, or order without sufficient funds; "conduct not becoming an officer and a gentleman"; and the omnibus "General Article," which covers "all disorders and neglects to the prejudice of good order and discipline in the armed forces, all conduct of a nature to bring discredit upon the armed forces, and crimes and offenses not capital, of which persons subject to this chapter may be guilty."[7] This code is far more stringent than U.S. civil codes because the norm of military service requires disciplined, mutually supporting efforts in the face of mortal danger.

Strictures added by international law and arms control further help ensure that the military remains within its legislated code. Indeed, the United States—highly sensitive about abuse of power, and keen to constrain those who command the weapons of war—over the years has taken extraordinary efforts through operational, organizational, legal, and moral means to ensure that military operations are carefully circumscribed.[8]

The vital importance of the code was captured by Lee Harris, when he wrote: "In the amplest perspective of world history, this is what distinguishes those handful of civilizations that have come to exercise inordinate military might: the code of honor that keeps in check their soldiers' hunger for more power. And it is the disintegration of this code of honor that dooms all human empires."[9]

Understanding this centrality offers an insight, for example, into why at military academies codes of honor have been instituted and why relatively minor infractions of conduct make national headlines. Commonplace on U.S. college campuses, and rarely reported in the news media, tales of religious coercion at the Air Force Academy, of chaining a female student to a urinal at the Naval Academy, and of hazing and cheating at West Point are often treated as front-page news. On the other hand, it also offers a partial explanation of why the U.S. military profession is held in such high esteem among U.S. citizens.

As a foundational element for military culture, the code underwrites its essentials: discipline, professional ethos, ceremony and etiquette, and cohesion and espirit.[10] Warriors need discipline to keep them within the bounds of the code, a professional ethos to provide direction and substance to the discipline, ceremony and etiquette to lend legitimacy based on history and tradition, and spirit to enable those bound by the code to perform with dedication and heroism.

Military culture, and especially Navy culture, encompasses a community bound together by trust. Trust must be as strong between leaders and subordinates as it is among equals in hierarchical organizations such as the military. Eventually trust begets loyalty, and all military officers recognize the importance and value of "loyalty up and down the line." Because the stakes include the ultimate—risking, or perhaps even sacrificing one's life—in military callings, trust and loyalty are pivotal to the coherence of military organizations. Thus, trust is important within the military because of the extraordinary demands that can be levied on its members, and trust is vital between the military and the state because of the pivotal role the military plays in the very viability of the state. In all instances, relationships involving trust and loyalty are annealed by the code.

An understanding of these sinews of military culture explains many things. For example, male bonding—a demonstrated effective method of implementing key elements of the culture—has a storied history in military organizations, and particularly in navies, which can be physically isolated from civil society for long periods of time, leading to the formation of tight-knit communities. To build trust and loyalty, military organizations employ discipline, and simultaneously promote a spontaneous willingness to follow orders to perform difficult and perhaps extremely unpleasant acts. Teamwork is fostered and nourished to focus energy and effort in order to fulfill mission requirements. Teamwork is founded on the equality of members of the team. It typified Horatio Nelson's "band of brothers" approach to naval warfare, and has been characteristic of the most professional of fighting units throughout history. Brothers form teams of equals, and join together to attain assigned objectives. Shared heroes and histories, as well as the sharing of a common fate, foster emulation and bonding.

The ocean environment demands an intimacy among shipmates that is dissimilar from landward societies. Ships are confining. They force close associations among shipmates. Ships are minimally manned, for additional crewmembers require space, food, and fresh water, which are all precious commodities aboard ship. There are no onboard reserve personnel or "supernumeraries" to stand the watch or operate the guns if one of the crew is sick or injured. The ship must be self-sustaining, and self-reliant. The environment is inherently hostile; teamwork is required merely to survive. Outside help when problems arise is rarely available. This engenders a greater feeling of community than exists in other military services. Because "the first principle of a seaman's outlook is the safety of the ship,"[11] fire aboard ship (and even more vitally aboard a submerged submarine) is high on the list of ultimate horrors: for if the fire doesn't kill, the sea certainly will. There is rarely a viable way to escape a burning ship, and no local fire department to call. So, everyone must be physically able to fight fires, and firefighting drills are a *daily* occurrence on warships. Everyone must be capable, for example, of wrestling a charged fire hose, running while toting a heavy fire extinguisher, and carrying a shipmate who is overcome by smoke from a burning compartment to a safe place.

Bonding derives from shared understandings, a coherent organization that encourages it, and the confidence-building effects of first-rate equipment. Anything that undermines the bonding of warriors or the code under which they fight can result in grave damage to the institution. It is this bonding that renders the whole greater, much greater, than the sum of its parts. Bonding is what inspires fighting men who have been seriously wounded to want to return to their unit, to fight with greater force and fierceness in the thick of battle, and to form lifelong associations after service in the military. Speaking on behalf of the British military, one of its officers set forth the premise in the starkest of terms: "We have a duty to be different, to maintain the ethos that makes people fight well. If we lose that ethos . . . then somewhere downstream, we shall lose a war."[12]

Common understandings within Navy Strategic Culture are found in concepts. Naval officers are comfortable with concepts, as distinguished from definitions.

"Define your terms, sir!" How many times one hears this in military staff environments. Yet, anyone who has attempted this exercise realizes that many important notions are either not precisely definable, or are defined for purposes other than conveying truth. Excessive attention to the formulation of definitions comes often at the expense of clearer questions. Those who prepare and work on interagency tasks in the U.S. government arena vie constantly for authorship, acknowledging that the authors of policy wield the power. Composers write what they choose to write, and then others must make persuasive arguments in order to change what has been written. The original author has no comparable burden of evidence or logic. As a consequence, one proven method to attack the person with the pen is to require definitions so that all participants find themselves, at least ostensibly, on the same page.

Yet, in a real sense, definitions represent the end of thinking about a subject rather than its beginning. Moreover, they are notoriously difficult to agree upon, for in an arena of contrasting views and positions they often appear designed, no matter how benign or well intentioned, to favor the preparer of the definition. Of equal importance, however, is that definitions are as rigid, concrete, and inflexible as their authors can make them. They imply, and are intended to imply, precision.

Concepts stand in stark contrast to definitions. Concepts are abstract, indistinct, and malleable. Concepts are not unbounded, however. They perform within conceptual limits, yet those limits allow the flexibility for the concept to change shape and to grow. Consequently, concepts and definitions stand opposed, as enemies. To define a concept is to kill it, much the same as to conceptualize about a definition is to undermine its meaning and value. For Navy personnel in the process of articulating a concept to be told to "define their terms" is to ask them to steer their ship directly at the shore. Those who think about naval matters are, for the most part, entirely at ease with concepts and quite uninterested in definitions. Hammering out definitions looks much like a futile exercise without much in the way of payoff, for definitions are often ignored when the chips are down and expediency—or, at the extreme, life and death—rules.

How did naval officers become more comfortable with concepts than definitions? By the ways in which strategic culture is shaped: study, understanding, and experience. One of Napoleon's maxims claims: "An admiral commanding a fleet and a general commanding an army are men who need different qualities. One is born with the qualities proper to command an army, while the necessary qualities to command a fleet are acquired only by experience."[13] Study and experience are both guided by leadership, and both are acutely sensitive to context.

Context is the centerpiece to understanding Navy strategic culture. The study of naval affairs and of naval history takes place in a setting in which leadership provides the beacons, but events and how they were resolved in the prevailing environmental, social, economic, and political contexts constitutes the learning process. One cannot understand how U.S. naval forces resisted some two thousand ferocious kamikaze attacks in the battle for Okinawa alone without contextual knowledge of how that menace evolved and what operational actions were taken to combat it. The heroic actions of the sailors of Task Unit 77.4.3, "Taffy 3" in the Battle off Samar in October 1944 cannot be internalized fully by those who have never "gone down to the sea in ships."

Likewise, to sail the open ocean in a ship and for weeks see nothing familiar—no buildings, no schools, no stores, no traffic lights, no advertisements, no automobiles—delivers a psychological impact. It is this lack of contextual familiarity that makes the environment so unique. It causes one to interact with the environment, and with one's shipmates, in a distinctive way. When the weather turns nasty, shelters from the storm tend to be few and far between; there's no 911 to dial for help. When landlubbers fly to Cleveland and the weather turns sour, they can instead, and easily, divert to Detroit, Columbus, Pittsburgh, or Indianapolis. When a naval aviator goes aloft from an aircraft carrier and the weather deteriorates, however, he rarely has a "bingo" (alternate) field to which to divert. Alternate airfields are as rare at sea as they are copious on land. These environmental facts discipline the thinking of those who experience them.

Embarking in a ship and putting to sea is to enter another world. It engenders a unique, rather peculiar language, which establishes a certain psychological distance from those who are land-bound. It enjoys its own special vistas; and, indeed, operates largely under its own rules.

The submariner operating a nuclear-powered submarine demonstrates the ideal. His environment is the most different from any other—especially that familiar "home" environment. In addition, he is deprived of several of his senses to understand and detect things in his environment. He has no sense of smell or touch with respect to his environment. His sense of sight is drastically limited. His most important and acute sense is that of hearing, and that is indirect—and artificially enhanced—only. His world is very narrow and, truly, very small. His concerns are farther down on the Maslovian hierarchy than others who go to sea in ships. He is much closer to survival requirements than others. Here, truly, the environment will destroy him if he lets his guard down for an instant. The environment will not kill the surface officer or the aviator unless they are very careless. The environment will slay the submariner if he is at all careless.

Navies operate in unique oceanic settings—ones experienced by 5 percent or fewer of the world's citizens. Not much there is well defined, but concepts abound. Take the horizon, for instance. It is, as everyone knows, where the earth's surface and sky appear to meet. But, "you can't get there from here" or anywhere else. And the distance to the horizon changes as height of eye (above the earth's surface) increases. It also changes with changes in ambient light or weather, and with the interposition of physical objects. People talk about "broadening horizons" and "creating new horizons." Naval officers know that the horizon is an ever-changing concept, and that "over the horizon" means "not known" or, "difficult to know." Indeed: "Now bring me that horizon!" are the last words of the motion picture *Pirates of the Caribbean.*

When one overlays combat operations on open-ocean settings, the fraction of "practitioners," those with experience, falls even more dramatically. The seas form the "great commons" over, on, and under which naval affairs take place. What happens to those who sail and fight on, over, and under the seas—and what they read about those who have—conditions the way they approach and seek to resolve current and future issues. Those contextual factors will be addressed in detail in the next chapter, but suffice it here to assert that they are real, they are different from those that have an impact on other services, and that they have deep and lasting effect.

To understand naval operations requires an appreciation of theory and thought patterns that require a unique vocabulary. Nautical terms are encrusted with brine: from "galley" to "gunwale," from "scuttlebutt" to "swab." Specialized terms are employed to describe operations at sea as well, which in themselves suggest important differences to operations on land, and that those variations must be studied to be understood. A few samples from a rich menu should suffice: convoy, fleet-in-being, commerce or cruiser warfare (*guerre de course*), fortress fleet, crossing the T, broadside.

Institutionally specific language separates different professions, but few as completely as for seafarers. In some of the most basic terminology, ships "still have bulkheads, not walls; cabins or compartments, not rooms (except in composition like wardroom, storeroom, etc.); overheads, not ceilings, and decks, not floors."[14]

Operating in ships outside the sight of land, day in and day out, constantly in motion without respite from the environment, incessantly vigilant for oceanic traffic and foul weather—and in time of hostilities to surface, air, space, and submarine surveillance and attack—naval personnel share bonds totally unfamiliar to landsmen. "We're all in the same boat," is true every hour of every day, of every week, of every month when the ship is at sea. It means, and the crew all know, that they share a common fate: that they must work, eat, fight, laugh, cry, and even die together. While this can be true of air crews also, they experience the intensity of it during their mission time only. For seamen, facing constant danger is not a passing task to be endured for a couple of hours, but a pervasive way of life.

The tradition of the Navy is one of independence in operations, success in battle distinguished by courage and sacrifice, and of technological excellence. Indeed, technology is required not only for survival, but for enhancement of all the relevant senses in an oceanic environment. Thus, the Navy "mans the equipment" while the Army "equips the man." When in Pentagon discussions one speaks of "force structure," the Navy thinks of ships and aircraft, while the Army thinks of divisions of soldiers.

This is of more than passing importance, for in land warfare one of the keys to successful combat is to maintain strong unit cohesion and to seek to disrupt the unity and coordination of adversaries, rendering them vulnerable to piecemeal defeat. At sea and in the air, the equipment—ships and aircraft—bind the fighters together; they cannot be defeated by psychological means alone. Their platforms must be destroyed or neutralized in order to defeat them. Because air and naval forces are highly mobile, and because they must be physically damaged or destroyed to be rendered *hors de combat*, they can, on occasion, opt to refuse battle. This is especially true at sea if the adversary does not have effective weapons to attack maneuvering naval forces on, over, and under the sea.

The skills that blue-water navies must acquire and perfect include conducting reconnaissance, surveillance, tracking, and weapon employment in all weather conditions and conducting a wide variety of combat operations—anti-submarine, antiwar, antisurface, and striking targets ashore—including amphibious landings; mine countermeasures; special operations, and counter-reconnaissance and surveillance—electronic warfare, operations security, and deception. Because modern navies rarely engage in hand-to-hand combat, it is less necessary to demonize the enemy. The enemy is a "concept," one who can kill you, but not necessarily on a personal *mano a mano* basis. Present-day sailors do not often experience the bloody carnage, the furor, stench, and clatter of combat, nor are they constantly under the unblinking eye of the modern media. They are divorced from those factors by their environment.

The contrast with the following list of "specific operating skills" compiled for the Army by the RAND Corporation is striking:

- Facility in joint and combined arms operations
- Dealing with civilian populations
- Force protection
- Operations in urban or restricted terrain
- Understanding the enemy situation
- Using technology for situational awareness
- Integrating coalition forces
- Interacting with media.[15]

Such differences are bound to stimulate contrasting, even clashing, views about how military conflict is to be averted and, if necessary, fought and won.

The dominance of context, especially environmental context, in the conceptual thought of U.S. Navy officers helps to explain the Navy's long-standing aversion to written doctrine. As an articulation of "how forces prefer to fight," doctrine nevertheless is non-specific as to place, opponent, or time. To the commander, doctrine is essentially context-free. Adding context, applying doctrine transforms it into tactics, operational art, and strategy. The commander acts as the creative force, mindful of the guidelines of doctrine but adapting them to fit the specific context of his or her mission. In this way tactics, not doctrine, addresses the question: "How do I employ forces in battle against adversary X in location Y at time Z with the weapons, organization, and information available to me?" Generally speaking, contextual application of leadership, foresight, anticipation, adaptation, prudent risk, courage, and perhaps a bit of luck, transforms doctrine into global, theater, or battlefield action.

Doctrine tends to be backward looking but forward leaning. It must be grounded in experience, carefully selecting those lines of effort that have been successful in the past, while rejecting those that resulted in failure or inconclusive results. It must exercise care to learn the right lessons, and mechanisms have been put in place to identify and to learn the appropriate lessons from past military actions. At the same time, it seeks to avoid the criticism that "the military is always fighting the last war." So, doctrine attempts to draw on past knowledge in order to lean forward and anticipate how to apply military force in the future. It seeks to determine the preferred way to employ technologies that have never been used in operational situations. The same is true of many weapon systems where there is no operational experience on which to base doctrine for their use. Nevertheless, doctrine for them must be developed, and this offers another insight into how doctrine must be "forward leaning."

Nor can doctrine act as a script, or menu for military operations. It guides actions; it does not prescribe them. Since it is essentially context-free, doctrine cannot account for the fog or friction of warfare. The commander must use doctrine intelligently—neither dismissing it as irrelevant to his current situation nor following it rigidly. Instead, the commander must employ military art to apply doctrine and transform it into strategy, operational art, and tactics.

Doctrine can inhibit imaginative thinking and battlespace adaptability. Julian Corbett, a seminal author on naval strategy, wrote: "Nothing is so dangerous in the study of war as to permit maxims to become a substitute for judgment.[16] Navy strategists look upon written doctrine as "maxims" in this sense, and are wholly uncomfortable with it. To the naval strategist, the combination of definitions and doctrine becomes rather toxic. Doctrinal thinking argues thus: "Define a point. Define two more points. Connect the points with lines. Color inside the lines." To this approach Navy strategists are very likely to say: "I don't even like how you've defined your points. And "color inside the lines—you're kidding, right?" In the motion picture *Patton*, the great general flamboyantly highlights the distinction being drawn here when he, from a point high above the North African battlefield upon which scores of tanks are still smoldering, exclaims: "Rommel, you magnificent bastard, I read your book!"

One often hears that U.S. Navy culture is excessively Mahanian: that it is wedded firmly to the ideas of Mahan, and his focus on battle fleets and great battles at sea—especially his classic, *Influence of Sea Power upon History*. Pundits have written that the Navy must look forward and not back, exorcising the ghost of Mahan, being wary of "Mahanoia" and "Mahanofrenzia." Those making such assertions intend to tar the Navy with brushes labeled "hidebound," "antediluvian," "doctrinaire," "rigid," or prone to "steering by its wake."

It is true that Mahan had many insights on strategic culture in general, and relevant to Navy Strategic Culture in particular, and that many of them apply today. Mahan was, after all, "the first to conceptualize the role of sea power in the human affairs of his time, rescuing from the forgotten historical record the relevant details of how victory at sea was arrived at and what benefits devolved to the victor at sea."[17] Of course, Mahan's ideas were circumscribed by the era in which he wrote—submarines and aircraft, for example, played no part in his strategic musings. Yet, there was much Mahan had to say that remains of relevance if one is to understand today's Navy Strategic Culture.

In the final analysis, service strategic culture involves establishing a rich, complex system of trust, loyalty, dedication, espirit, and commitment. This is accomplished in the first instance by the swearing of an oath, one that establishes the boundaries within which the professional officer must act. For naval officers, their own particular strategic culture is shaped first by their education—at the U.S. Naval Academy, in Naval Reserve Officer Training, or in Officer Candidate School—and then by the environment in which they work. Their approach to the application of naval force is tempered along the way with an appreciation of institutional history, experience at sea, focused training, and interaction among other services and members of the services of other countries. The acculturation is a daily and constant process, for, "We are born with instinct, but in matters of intuition, we are lifetime learners."[18]

An appreciation for history is particularly pertinent, for opportunities to participate in naval combat are scarce. As the great historian B. H. Liddell Hart observed: "Direct experience is inherently too limited to form an adequate foundation either for theory or for application. At best it produces an atmosphere that is of value in drying and hardening the structure of thought. The great value of indirect experience lies in its greater variety and extent."[19]

Owing to systemic feedback, the culture is constantly and relentlessly tested and honed—day after day. Yet, changes tend to be slow and marginal, for the core concepts are long-standing, deeply rooted, and proven.

The instruments of culture are not only resilient, they demonstrate intrinsic value. Importantly, "The human ability to learn from experience and nature, so slighted in current humanistic theory, is not merely an object of cultural transmission, let alone of social control, but an evolutionary triumph of the species, indeed, a triumph on which our future ultimately depends."[20] This means clearly that the study and understanding of culture in general, and for the purposes of this book, Navy Strategic Culture in particular, is a valuable undertaking.

Strategic culture is rooted in history, common understandings, and shared experience. Navy Strategic Culture must be viewed largely as the by-product of sailors operating communally on, over, or under the open ocean for extended periods of time. It manifests itself in the way navy officers approach problems, and how they apply what they have extracted from history, training, and experience.

Whether or not culture determines actions, it is enough to say that it strongly influences both attitudes and behaviors. The great Thucydides wrote: "It must be kept in mind that seamanship, just like anything else, is a matter of art, and will not admit of being taken up occasionally as an occupation for times of leisure; on the contrary, it is so exacting as to leave leisure for nothing else."[21]

CHAPTER 3

The Maritime Context

Scientific studies have long noted the power of external symbols to influence memory, and hence, action. "External symbols give us stable, permanent, virtually unlimited memory records that are infinitely reformattable and more easily displayed to awareness than the brain's more limited biological memory records."[1] The context, the oceanic environment in its many manifestations, provides a set of symbols and patterns that powerfully shape the culture of those who sail the seven seas. These symbols and patterns establish a context markedly different from that experienced on terra firma. It is different enough, indeed so unique, that it merits study for what it can reveal about how context conditions Navy Strategic Culture and, as a consequence, affects military maritime operations. This complex oceanic setting exhibits physical, political, legal, and economic properties.

Physical

First, foremost, and most determinative is the physical environment. Size is the physical characteristic that stands out. It is well known, for example, that water covers about three-quarters of the earth's surface, and a high percentage of that is navigable by ships. Indeed, the Pacific Ocean alone is larger than all the land masses of the world combined. The average depth of the water in the oceans is over twelve thousand feet, and the oceans contain 97 percent of the world's water.

"It is estimated," furthermore, "that if all the inequalities in the height of land and depressions of the sea bottom were to be leveled off, the entire earth would be covered by a uniform layer of water one and a half miles in depth."[2]

At the surface of the earth atmospheric pressure is 14.7 pounds per square inch. Beneath the seas the pressure increases with the depth of the water. This limits the operational depth of submarines to about twelve hundred feet, where the pressure on the outer hull of the submarine is in excess of five hundred pounds per square inch. Recalling that the average depth of the oceans is ten times this value, one can understand that submarines can use effectively only 10 percent of the ocean's depth, but that still offers them a great volume in which to operate—a total of about 32 million cubic miles.

Because they are so greatly attenuated, electromagnetic and optical energy are not useful for reconnaissance under water. This leaves sound as the preferred medium for underwater reconnaissance and surveillance. Sound travels at about four times the speed in water as it does in air; and, under certain conditions, unlike in the atmosphere, in water sound can travel very long distances—thousands of miles, in fact. The speed of sound in water varies directly with temperature, pressure, and salinity. That is, the higher the temperature, pressure, or salinity, the higher the speed of sound in water. Deep sound channels in the ocean can transmit sound rapidly to distant receptors. While salinity is nearly constant in the open oceans, it can vary considerably in the coastal zone, especially in the vicinity of an estuary.

Knowledge of the sound, temperature, and salinity profiles of the water column, as well as the configuration and composition of the bottom, and considerations of ambient noise permit estimates to be made about sound propagation as a function of depth. This is important for submarine, anti-submarine, and mine hunting operations.

The depth of the water conveys important operational information. For example, sea mines can be laid effectively only in relatively shallow water, which is found, for the most part, near land. The shallow North Sea, which has a average depth of some three hundred feet, was the scene of significant mine warfare in World Wars I and II. In World War I about seventy thousand mines were laid in the "North Sea Barrage," and in World War II nearly six hundred thousand sea mines were emplaced in the European theater by both sides. As another example, there is a minimum depth of water in which submarines can submerge and minimum depths into which homing torpedoes (ones that have self-contained sensors and guidance mechanisms) can be employed. Submarines benefit from deep water because the deeper a submarine dives the faster it can go without its screw propeller causing telltale cavitation that can be detected by passive sonars.

The surface of the sea is easily penetrable—unlike the land. Consequently, ships can sink, and so can people, and sinking tends to be fatal to both. A seaman's first, visceral battle is with the environment. It is always trying to kill him. This highlights the fact that technology is essential to seafarers. Aircraft, because of their high speed, and submarines, because they can submerge, are better able to master their environments and avoid bad weather than surface ships. But all are highly dependent on technology.

"War at sea," wrote Victor Davis Hanson, "is a primordial killer of men, in which the ocean itself can wipe out thousands without the aid of either man or his weapons. At Salamis most died from water in their lungs, not steel in their bodies."[3] The battle to stay afloat, and not succumb to the elements, is constant for the seafarer. He must first win this battle, or at least manage it successfully, so that he might then address human adversaries.

Ships do not travel fast. As a consequence, great distances translate into extended travel times. To sail by surface ship from San Francisco to Tokyo takes over twelve days at an average speed of fifteen knots; Boston to Gibraltar, slightly over eight days. "Knots" are nautical miles per hour. A nautical mile is 6,076 feet in length, whereas a "land" or "statute" mile is 5,280 feet, or 86.9 percent of a nautical mile. Put another way, a nautical mile per hour equals 1.15 statute miles per hour. Global—air and sea—distances are measured in nautical miles because the definition of a nautical mile is "one minute of latitude," so distances are easily measured on maps and charts by using the latitude scale at the latitude at which the measurement is taken. Nuclear submarines transit at higher speeds, because their endurance is not limited by fuel expenditure, and because they are submersed in water they can adopt a hydrodynamic shape that permits higher speeds. Most of the transit time for surface ships, and virtually all of it for submarines, of course, takes place beyond the sight of land.

In the era of sailing ships, the time a warship could remain at sea without replenishment was limited by sustenance for and health of the crew. When steam-powered warships replaced sail-powered ones, the limiting factor became fuel for the ship. It needed to be replenished more often than provisions for the crew. With nuclear-powered warships, endurance changed back to being limited by the crew's physical welfare.

Geographic size of the world's oceans is important because it provides the key aspect of security: sanctuary. To be secure is to feel safe, and militarily to have the sense that one can operate both without hindrance and safely. Sanctuary can be imparted by active and passive defenses—antiaircraft missiles and armor plate respectively, for example. It can also be provided by making its beneficiary difficult to find. The mere size of the oceans provides a near-global hiding place for ships—especially for submarines.

The open oceans are not only *aqua incognita*, but except for the sun, stars, and planets, they are unmarred by natural navigational assistance. Before the advent of electronic navigational aids such as LORAN and OMEGA, and even more recently satellite navigation, sailors could determine their location on the open seas with precision only when the sun, stars, and planets were visible, and then only after the invention of accurate, seaworthy timepieces.[4]

The sheer size of the water areas of the earth, complemented by the earth's round shape (which will be discussed later in this chapter), the absence of daylight approximately half the time, and the vagaries of weather mean that sanctuary is widely available for ships. The fact that sovereignty—political control—over open ocean areas is imposed very differently than over land counts as another factor in favor of sanctuary for ships.

In all ocean and sea areas, winds, waves, precipitation, ice, cloudiness, currents, and air temperature affect the employment of ships, aircraft, and the use of their sensors and weapons. These factors are generally more pronounced in the coastal areas and narrow seas, such as straits and passages. Ocean currents greatly affect Earth's climate by transferring warm or cold air and precipitation to coastal regions, where they may be carried inland by winds.

Some of the most dramatic forms of weather are spawned over the oceans: tropical cyclones (also called "typhoons" or "hurricanes" depending upon where the particular system forms). The passage of severe storms add to the roughness of the sea. The sea's roughness can be reduced in the winter by ice cover in high latitudes and, for example, in many parts of enclosed waters such as the Baltic Sea. Rough seas prevail mostly in the fall and are less frequent in the spring. High sea states influence the speed of all surface ships and affect the crew's comfort, cause fatigue, and interfere with flight operations and all other topside activity. Sea states are numbered from zero to nine. A sea state of five, for example, would have a wind speed of twenty-one to twenty-five knots, and a significant wave height of from eight to twelve feet. A wind of thirty knots and significant wave height of seventeen feet would describe a sea state of six. (A table of sea states with corresponding wind speed, significant wave height, significant range of periods, average periods, and average length of waves can be found at http://www.eustis.army.mil/weather/weather_products/seastate.htm [August 17, 2008].)

The employment of small surface combatants in the Baltic Sea is seriously hampered for about two months of the year because of strong winds and high seas rather than ice. The North Sea is characterized by violent northwestern storms in winter months that make navigation along the southeastern coast dangerous, especially along the Jutland Peninsula. Rain and fog are frequent occurrences in all seasons in the Baltic and the North Sea. Narrow seas at lower geographic latitudes—such as the Arabian Gulf and the Red Sea—however, are more favorable to the employment of small surface ships, owing to the generally less extreme weather conditions. In the Arabian Gulf thunderstorms and fogs are rare, although dust storms and haze occur frequently in the summer months. The most common and strongest wind, which blows from the northwest and west-northwest, seldom reaches a velocity of twenty-five knots and rarely thirty-five knots.

Combat employment of conventional-hulled combat craft becomes difficult or even impossible at a sea state of five or higher. Then the craft's speed must be drastically reduced to avoid structural damage caused by waves, and the discomfort and stress imposed on the ship's crew.

The open ocean is not only vast, but it is a non-linear space: there are no lines, no demarcations, no established places to organize or structure movement. When ships can virtually disappear on the open oceans for weeks or months at a time, when they can literally pass within shouting distance of one another at night or in a fog totally unaware of the other's presence, where no roads or valleys or other geographic formations serve to guide or channel their movement, it becomes more and more evident that the key operational problem for warfare on the open oceans is to find the opposing force. If an adversary naval force can be located in such an environment, then tracking it, classifying it, and ultimately targeting it with weapons appear much less difficult in comparison.

Spatial non-linearity means also that in the open ocean there is no obvious threat direction from which attack by submarines, surface ships, or sea-based aircraft might originate. There is no clearly defined threat axis. Against land-based aircraft, the potential threat direction can be estimated at long ranges, but it becomes more and more circular as the distance to adversary airfields decreases. Antiship cruise missiles and torpedoes can be launched from surface ships, aircraft, and submarines. Unlike structured land warfare, an attack against a surface naval force from a capable adversary can originate from any direction—even from under the seas—at just about any time.

The corollary to the difficulty in locating an adversary is that with the knowledge that findability is key, one does everything one can to maintain stealth—to preserve one's own sanctuary. One conducts random movements, concealing actual destinations if possible. One paints combatant ships gray so that they meld with the background. One transmits very judiciously—or not at all—on omnidirectional radios, radars, or sonars so that counterdetection opportunities are significantly reduced. One seeks to reduce emissions into the environment that might aid detection by an opposing force: heat, noise, and smoke in the daytime and heat, noise, and light at night. (Even gunpowder comes in two varieties: smokeless for daytime and flashless for nighttime use.) One develops and deploys submarines, the stealthiest of ships, and equips and operates them to be very quiet.

If detection occurs, then a reduction in the clues that give away the ship's identity becomes paramount. Combatant ships no longer operate in symmetrical circular formation with the most valuable ship at the center as the bull's-eye. Random formations, at one time called "haystacks," are the norm. In wartime ships have been painted with camouflage paint to confound an adversary's visual cues about their type, size, or course and speed. Even if they can be detected, the identity of submerged submarines can be extremely difficult to determine, especially in view of the fact that submarines of the same manufacture are operated by different countries.

If efforts to avoid or nullify detection, tracking, and classification by a reconnoitering force all fail, then one must turn to passive and active defenses as the final attempt to provide sanctuary against attack. These are key facts about warfare in the naval realm. They have been true since ships first sailed over the horizon, and have exerted seminal influence over strategic culture, and, accordingly, over naval policy and strategy.

The open oceanic domain is essentially featureless, and the curvature of the earth further restricts the ability of those on surface ships to view their surroundings. What is called "direct line-of sight" is in actuality very limited. For example, if one were to stand on the deck of a ship so that one's eye was fifty feet above the water, the distance to the horizon would be just 8.1 nautical miles. If the height of eye is one hundred feet, the horizon is at 11.4 nautical miles.

Electromagnetic signals, like radar, add about 15 percent to the visual line-of-sight range owing to the curvature or refraction of the signal. So, a radar antenna one hundred feet high can, theoretically, detect something on the sea surface at a range of slightly over thirteen nautical miles (11.4 plus 15 percent). These ranges must be considered nominal, for they are affected by atmospheric conditions; by the material condition of the radar and its frequency; and by the target's size, aspect (orientation with respect to the radar signal), radar reflectivity, and height. Of course, the curvature of the earth's surface also limits the ability of those on land to observe what is happening at sea. For example, the distance to the horizon for a person six feet tall standing at sea level is just 2.8 nautical miles.

To extend the line-of-sight, one must elevate the receptor to high ground, or locate it on aircraft or on a satellite in space. Of all the human sensors, only eyesight is of utility in the atmosphere at sea, and it can be significantly impaired by darkness or bad weather. Consequently, augmentation of human sensors is required by electromagnetic or electro-optical means.

Of interest also, is the fact that whether one is using electro-optical or electromagnetic sensors, unless the target is close to the observer, (or is in orbit) it will be found low on the horizon. Table 1 demonstrates this fact. Accordingly, if an aerial target is one thousand feet high in the sky, it will be above one degree in elevation only once it reaches 8.87 miles from the observer. If it is at an altitude of ten thousand feet, it will be above one degree when it is at a range of 64.7 miles. Of course, if the aircraft or missile maintains a constant altitude as it approaches, it will initially "pop up," and then appear to climb higher and higher in the sky. Finding aerial targets on the open oceans as far from the observer as possible, consequently, always takes place at very low angles of elevation.

From Table 1 it can be observed that very low-flying aircraft or missiles, say at an altitude of one hundred feet, will "pop up" over the horizon (at about one-sixth of a degree in elevation) less than five miles away. If the attacker's speed is five hundred knots, it will be about thirty-five seconds from impact. This affords precious little time to react. Of course, the attacker must have a very good idea of where its target is so that its weapon systems can acquire it. This is all part of the "hiders and finders" struggle that occupies the central core of naval combat.

Long-range shipborne radars must be able to scan at very low angles to find targets as far from the ship as possible. Sea return, or "clutter," from the sea surface degrades radar capability, and high sea states will degrade it further.

Clearly, the best way to extend one's detection capability is to place the sensor aloft—the higher, the better—satellites offering the ideal in this regard. From a perch in the sky, one can see "over the horizon" and thereby greatly increase reconnaissance and surveillance capability. When aircraft first went to sea on ships, they were used exclusively for this purpose, and today, when they are available, they serve a similar function. The benefits of long-range reconnaissance and surveillance resulting in early warning of approach by other forces are offset somewhat by the possibility of counterdetection.

Long-range search and detection can be undertaken through the air with active sensors such as radar, or underwater by the transmission of sound signals—i.e., active sonar. If another platform—a ship, submarine, aircraft, satellite, or

Table 1. Height vs. Elevation Angle vs. Distance

Height of Target	Elevation Angle	Distance to Target
50 feet	1/6° (10 min.)	2.55 mi
	1/2° (30 min.)	.93 mi
	1°	940 yds
	2°	480 yds
	4.4°	200 yds
100 feet	1/6° (10 min.)	4.72 mi
	1/2° (30 min.)	1.83 mi
	1°	1860 yds
	2°	940 yds
	5°	380 yds
	9°	200 yds
1,000 feet	1/6° (10 min.)	26.6 mi
	1/2° (30 min.)	15.48 mi
	1°	8.87 mi
	2°	4.64 mi
	5°	1.88 mi
	10°	.93 mi
	20°	960 yds
10,000 feet	1/6° (10 min.)	104.4 mi
	1/2° (30 min.)	85.5 mi
	1°	64.7 mi
	2°	41.2 mi
	5°	18.3 mi
	10°	9.3 mi
	20°	4.5 mi

Derived from Nathianel Bowditch, *American Practical Navigator: An Epitome of Navigation.* H.O. Pub. No. 9 (Washington, D.C.: Government Printing Office, 1958), Table 9. Distances in nautical miles.

missile—is emitting energy, that energy can be detected by passive sensors. The eye is a good example of a passive sensor; unaided, however, it is of limited utility to the seaman. The advantages of passive sensors are that they cannot be counterdetected, and that they can detect signals from active emitters at a greater distance than the active emitter can detect them. For example, a radar signal can be counterdetected about twice as far as it can detect targets. This is because the active signal is reflected from the target, and it must travel two ways to be received at the transmitting site. A receiver at a range beyond which the original signal has sufficient energy to return to its transmitter can receive the transmitted signal without the knowledge of the transmitting station. Not only can a radar signal be intercepted, it can be analyzed to determine the kind of radar from which it was transmitted—for example, a long-range search radar, a missile control radar, or a shorter-range surface search radar. This can assist in classification of the source of the signal: Is it a navigation radar from a merchant ship, or a height-finding radar from a guided missile cruiser?

Long-range, high frequency radio signals can be intercepted over great distances, and with multiple intercepting stations the location of the emitter can be determined. During World War II, for example, a system of high frequency detection and direction finding (HFDF) stations helped locate German submarines in the Atlantic that transmitted on their long-range radios.

Passive sensors—ones that do not emit signals and thus do not offer counter-detection possibilities to the adversary—are, obviously, to be preferred. The major disadvantage of a passive system is that it requires a "cooperative" target—one that will emit a signal that can be intercepted. The best kind of passive systems are those that can intercept signals that a target platform is obliged to radiate—for example, engine noise by a submarine or infrared signals from the hot engine or generator exhausts of surface ships.

Submarines can counterdetect active sonar transmissions at very long ranges, and as a consequence use their own active sonars only very sparingly. Nuclear-powered submarines operate steam-driven turbines for propulsion and generating power; and, even though they employ impressive technologies for noise suppression, they are not as quiet as a battery-powered submarine operating on its battery. The disadvantage of the battery-powered submarine is that the battery must be charged often, and high speeds can be sustained for only short periods of time before the battery is depleted. Charging the battery involves operating an internal combustion engine, which increases radiated, detectable noise.

Passive sensors are limited also by line of sight, weather, and ambiguity concerning distance to the intercepted signal. Passive sonar or electromagnetic intercepts can be made over great distances, but determining the range to the emitting force can be difficult.

This extended discussion of sensors underscores the attention and concern the at-sea naval force places on detection and preventing counterdetection. Darkness and weather help ensure that the notion of ships passing in the night unaware of the presence of one another is not confined to past centuries or sailing ships. Thus, while the environment offers no topographical features to provide cover from observation, the vastness of the oceans, the curvature of the earth, and natural phenomena make finding surface ships—not to mention submarines, which have an additional dimension in which to operate—an arduous chore. This is why the preponderance of at-sea battles throughout history took place within the sight of land, and with the opponents in visual range of one another. It cannot be too greatly emphasized that the major concern for modern naval warfare is that of finding the adversary, and its corollary: Don't get detected!

The final topic to be addressed with regard to the physical context of the maritime arena is that of navigation. In the non-linear, featureless at-sea environment, there are precious few navigational references. Mostly they exist in the sky. This offers another reason why John Masefield observed that "men in a ship are always looking up, and men ashore are usually looking down."[5] Mariners, to be sure, must have the utmost respect for Mother Nature. It builds both confidence and humility.

John Masefield was referring primarily to concern of seamen for weather; but maritime navigational help comes from the sun and moon, the planets, the stars, electronic navigational aids and, most recently, navigational satellites. The art of navigation is one of the prime skills of a naval officer. What are its tools? Sextant, chronometer, magnetic compass, lead line and fathometer, stadimeter, pitometer log, alidade, and navigation charts: important to measure or calculate water depth, current, tides, azimuths, and distances. Most of these are completely unknown to landsmen. Only since the Global Positioning System became fully operational does the mariner have an all-weather, accurate, reliable source of twenty-four-hour-a-day navigational information. The impact, and effect, on naval operations from this fact cannot be overestimated.

Political

As late as the 1940s, school textbooks on geography portrayed parts of central Africa as "unexplored." The high seas are distinctive not because they are unexplored, but because people do not live there and, as a consequence, tend to be unfamiliar with them. They are *aqua incognita* as much as central Africa was *terra incognita* to schoolchildren over six decades ago. From the earliest times, starting in the Bronze Age, seafarers put eyes on the bows of ships, for sailors were fearful not only of sailing beyond the edge of a the flat world into the abyss, but of torrential weather, sea monsters, enchanting sirens, perpetual darkness, and whatever else their imaginations might conjure. Even today the forwardmost part is called the "eyes" of the ship.

Since that time over six decades ago, the "unexplored" part of Africa has been brought under some degree of political control, however nominal. In comparison, the open oceans of the world are, and have always been, essentially politically uncontrolled. Land areas are divided up and administered by a large number of state and international political organizations. On the other hand, "With a few insignificant exceptions—salt lakes rather than seas—[the seas] are all connected one with another. All seas, except in the case of circumpolar ice, are navigable. A reliable ship, competently manned, adequately stored, and equipped with means of finding the way, can in time reach any country in the world which has a sea coast, and can return whence it came."[6]

Whereas controls over visitors from other countries are imposed on land, and permission must be obtained for overflight of another international state, the open seas are available for use by all. This near absence of political control means that the legal regimen for the high seas tends to be modest in scope. The politically uncontrolled nature of the oceans is the second of the characteristics that makes the maritime environment unique.

Where people live they organize themselves politically. Political contexts, likewise, are the breeding grounds for war.[7] Not since the early eighteenth century have major wars been undertaken with control of sea areas or maritime trade as their primary objective. Wars over the past two centuries have originated from political causes on land. Consequently, combat at sea, as has been the wise counsel of many maritime strategists, is important only insofar as it affects the course and outcome of wars on land. To the present time—with the single exception of the very minor Cod Wars between the United Kingdom and Iceland in the 1950s and again in the 1970s—neither political nor economic stakes have existed that warranted warfare with control over oceanic territory or resources as their objective.

Geopolitically, the United States has been blessed with a secure location and friendly neighbors. Once its borders were fully secured at the end of the Mexican-American War, external threats to U.S. security have not arisen from landward. Direct defense of the homeland—discounting the minor concerns of Spanish attack on the southeastern United States at the end of the nineteenth century and the insignificant Japanese efforts against the U.S. West Coast early in World War II—had not been a key concern for U.S. national security until September 11,

2001. Faced with political anxieties about infiltration of terrorists and weapons of mass destruction into the homeland and the vulnerability of coastal targets such as liquid natural gas terminals, chemical and petroleum plants, nuclear power facilities, and possible missile attack on coastal cities, the Navy has had to make a fresh evaluation of the threat, and to acknowledge that it will be obliged to pay greatly increased attention to homeland defense in the future.

This will require some significant alterations in how the Navy operates and interfaces with other navies of the world, of which there are now about 110, about a two-thirds increase in their number over the past fifty years. The count is somewhat imprecise because many states maintain forces with maritime functions such as coastal policing, customs enforcement, and a broad array of coast guard operations that are not organized as navies. Former Chief of Naval Operations Adm. Michael Mullen took a major step in this direction by developing the notion of a "one thousand ship Navy," in which the navies and coast guards of the free world would voluntarily band together in loose coalition for mutual support in carrying out the new challenges of penetration of states and attack from seaward axes.

Looking outward from national homelands, the world's oceans are divided politically into two parts. The first includes internal waters, territorial seas, and archipelagic waters of sovereign states. Such national waters are subject to the territorial sovereignty of coastal states, with some restricted navigational rights reserved to the international community. The second part includes contiguous zones, waters of the exclusive economic zone, and the high seas. These are international waters in which all nations enjoy the high seas' freedoms of navigation and overflight.

Because the majority of the world's capital cities and state populations lie within two hundred miles of a coastline, the political influence of powerful naval forces operating adjacent to foreign shores tends to be magnified. While—since the advent of the twelve-nautical mile territorial sea—offshore presence is no longer "visible," it is certainly "tangible." A coastal state might not be able to see ships cruising off its coasts, but it will "feel" them. A widely dispersed presence mission by an oceangoing navy can serve as a warning to adversaries, an indication of support to allies, and a demonstration of resolve that cannot be ignored by neutrals.

In the previous two paragraphs the notions of the demarcation of sea areas into zones of sovereignty and control have been raised, and this provides an opening into a discussion of the legal context for naval operations, which, of course, is tied tightly to the political context.

Legal

Naval operations take place in areas that are, for the most part, not under the jurisdiction of any single nation. Accordingly, the links between naval operations and international law are long-standing and close.

Having no effective supervening authority, and a comparatively small body of either treaty or case law, the legal regime for the ocean areas of the world is understandably weak. This, significantly, is by design—at least for a venerable U.S. policy of "freedom of the seas."

The United States has long maintained that beyond its territorial waters, no state can unilaterally exert full sovereignty over sea areas. This means that all nations have free and open access to the oceans for navigation and transit, for fishing, and for laying of submarine cables on the ocean floor. The principle of freedom of the seas, *mare liberum,* was first articulated in the early seventeenth century by Dutch jurist Hugo Grotius, and has been espoused strongly by the United States from its very inception. It is based on the fact that the commercial value of the seas lies in the ability to conduct maritime transport over them, not on exploitation of the area they cover—or extraction of their resources, with the exception of fish. Other states have attempted to control parts of the open seas from commercial competition, but the United States has resolutely resisted such control from the earliest times of the republic.

In *peacetime,* two sources of formal international law govern the operations of maritime forces. First, the Convention on International Regulations for Preventing Collisions at Sea, 1972, ordinarily called either the "International Rules of the Road" or the "72 COLREGS [Collision at Sea Regulations]." These rules—which are mandatory for all U.S. flag ships, commercial or military—set forth instructions for displaying lights at night and shapes during daylight, and the responsibilities between vessels with respect to right of way when ships are in danger of collision. The International Rules of the Road also detail proper sound signals between vessels, and stipulate correct actions in situations of fog or restricted visibility. Naturally, seagoing naval officers must be fully knowledgeable with respect to these rules.

The second major source of formal international law is the United Nations Convention on the Law of the Sea (UNCLOS), 1982. Under this convention, coastal nations of the world have established zones in the sea areas adjacent to their coasts, and formulated the rights and responsibilities to be exercised within those zones.

The zone closest to the land is that of the Territorial Sea, which extends to a distance of twelve miles from the coast. Beyond that lie the Contiguous Zone (up to 24 miles from the coast) and the Exclusive Economic Zone (EEZ), reaching seaward up to two hundred nautical miles. The coastal state maintains territorial sovereignty—subject to the concept of innocent passage—in its Territorial Sea. In the Contiguous Zone the coastal state has rights and responsibilities enumerated in the convention having to do with the state's customs, immigration, fiscal, and sanitary regulations. In the EEZ, according to the convention, the coastal state enjoys sovereign rights for exploring and exploiting, conserving, and managing natural resources.

The UNCLOS also delineates rights having to do with the establishment of artificial islands, installations, and structures; maritime scientific research; and the protection and preservation of the maritime environment. Of greatest importance to the Navy are the navigational provisions of the convention, which cleave to the notion of the unfettered ability of all to use the high seas for navigation and overflight. Article 87 of the convention, "Freedom of the high seas," is explicit in this regard: "1. The high seas are open to all States, whether coastal or land-locked. Freedom of the high seas is exercised under the conditions laid down by this Convention and by other rules of international law." The convention contains provisions for transit of international straits and archipelagic waters, and for innocent passage of the Territorial Sea of other states. It also asserts the complete immunity of warships on the high seas "from jurisdiction of any state other than the flag state," and contains provisions for, inter alia, the suppression of piracy, prevention and punishment for those who would transport slaves over the oceans, and for the omnibus provision for masters of ships to "render assistance to any person found at sea in danger of being lost."

Among the lesser body of conventions relating to oceanic areas is the recent *Convention for the Suppression of Unlawful Acts Against the Safety of Maritime Navigation of 1988*, with its 2005 Protocol. This convention is part of a wide-ranging effort by the United Nations to counter international terrorism.

These sources of international law are "formal" because they are secured by international legal agreement. Complementing these are other secondary sources of international law, including customary law (a matter of recognizing the general and consistent practice among nations over time to be accepted by them as a matter of legal obligation), general principles of law recognized by civilized nations, judicial decisions, and the teachings of the most highly qualified scholars. These secondary sources, problematic as they are, have been recognized in the Statute of the International Court of Justice as sources to which courts may turn to determine the law in any particular case.

As of spring 2009, the United States is not a party to the UNCLOS, although the treaty has been signed and has been before the Senate for ratification since 1994. This has not been considered a matter of urgency, because the parts most important to naval operations—the navigational provisions—are considered by the United States to be resolved customary law. Customary law, in international parlance, becomes binding over time because all parties accept it. The way in which customary law is prevented from becoming binding is to demonstrate that it is not customary, or to challenge it.

Because of the reach of customary international law, and since global navigational freedoms are central to its naval operations, the United States has conducted an organized, deliberate program to challenge excessive national maritime claims and practices. This program, called the Freedom of Navigation Program, began in 1979—even before the UNCLOS was completed in 1982. The kinds of claims that are challenged by U.S. warships under this program include, for example, improper historic waters claims, improperly drawn baselines from which to measure the zones provided by law, territorial claims in excess of those provided by law, and impermissible restrictions on non-resource-related high seas' freedoms in the Exclusive Economic Zones. Thus, if a state asserts a claim under customary international law, it can become binding if there have been no challenges to it. That is the basis for the U.S. program.

At the extreme, exercise of the freedom of navigation can result in shots being fired, as was the case in 1981 in the Mediterranean Sea. It happens that Libya claims all waters south of 32-30 north latitude, the Gulf of Sidra, as Libyan internal waters. This was considered an excessive claim, and on August 19, 1981, during the conduct of Freedom of Navigation operations in the Gulf of Sidra, a Libyan fighter aircraft fired a missile at aircraft from the USS *Nimitz*. The U.S. F-14 fighters shot down two Libyan Su-22 jets in the course of that incident.

The United States is the only country of the world with such a focused, consistent approach to challenging excessive maritime claims and practices. That is true because it is the country with the greatest perceived need to make all the ocean areas of the world safe and free for maritime commerce, to secure the oceans for its potential future military operations, and to prevent ocean areas from being used by potential aggressors as a sanctuary. No other state has either the capability or the will to mount such an effort. Indeed, because the United States has undertaken these tasks, no others have needed to invest in them.

Wartime sources of international law with greatest impact on naval operations include the Hague conventions of 1907 that set forth rules for the laying of automatic submarine contact mines, bombardment by naval forces, restrictions with regard to the right of capture in naval war, and the rights and duties of neutral powers in naval war; the London Naval Protocol of 1936 in regard to the Operations of Submarines or Other War Vessels with Respect to Merchant Vessels; the Geneva Conventions of 1949 with their Additional Protocols of 1977; and the 1980 Convention on Certain Conventional Weapons.

Among the thorny issues in the Law of Armed Conflict are the rights and treatment of neutrals in the battlespace and the establishment of exclusion or defense zones in ocean areas. The former arises from the fact that, unlike war on land, an oceanic battlespace can contain many neutral parties, ones that enjoy, under the law, immunity from attack. The status of neutral ships can pivot on such questions as what their cargoes are, where their voyage originated, and what their destination is.

The very definition of "contraband" cargoes is highly debatable, and has traditionally been resolved by the stronger party's enforcement of its own definition:

> Contraband consists of goods destined for the enemy of a belligerent and that may be susceptible to use in armed conflict. Traditionally, contraband has been divided into two categories: absolute and conditional. Absolute contraband consisted of goods the character of which made it obvious that they were destined for use in armed conflict, such as munitions, weapons, uniforms, and the like. Conditional contraband consisted of goods equally susceptible to either peaceful or warlike purposes, such as foodstuffs, construction materials, and fuel. . . . The practice of belligerents during the Second World War collapsed the traditional distinction between absolute and conditional contraband. Because of the involvement of virtually the entire population in support of the war effort, the belligerents of both sides tended to exercise governmental control over all imports.[8]

Exclusion or defense zones were established in both the World Wars to define areas in which neutral shipping would be banned or put in jeopardy. During the Cold War, the "battle for the first salvo," in which Soviet ships would shadow U.S. aircraft carriers and other "high value targets," in order that they might have an opportunity to strike the first blow in wartime, resulted in some effort in ways to establish a "maritime exclusion area" around the most valuable assets in a crisis, or to enclose a geographic area such as was done by the British in the Falklands War of 1982.

The difference in the nature and application of legal regimes in maritime areas from those on land has left its mark on the minds of seamen. They recognize that the rules are few, generally based on common sense, and it is in the interest of all to comply with them. Where the legal regimes are most restrictive and forceful is in the conservation of the resources of the seas and in matters of the security of coastal states.

For the most part, the legal regimes for the seas encourage global mobility and action. Bernard Oxman, a scholar in maritime law, has written: "Both collective self-defense and collective security under the United Nations Charter, including enforcement, peacekeeping, and humanitarian operations, continue to rest on the assumption of global mobility, which means under current law that naval and air forces enjoy the freedom of the seas in EEZs as well as the concomitant right of transit passage through straits connecting the EEZs."[9]

The central importance of freedom of the seas was set forth most concisely by President Ronald Reagan: "Freedom to use the seas is our Nation's lifeblood. For that reason our Navy is designed to keep the sea lanes open worldwide. . . . Maritime superiority for us is a necessity. We must be able in time of emergency to venture in harm's way, controlling air, surface, and sub-surface areas to assure access to all the oceans of the world."[10]

Economic

Discussion of freedom of the seas leads directly into the economic context in the maritime arena, for freedom of navigation has long been touted as essential to the economic health of the United States and its allies. The world inventory of ocean-going ships now numbers about forty thousand, roughly half of which are at sea at any given time. Almost five thousand of the latter are transporting petroleum products from the oilfields of the world to fuel the economies of industrial states.

In fact, over 80 percent of the world's trade, by volume, is carried in ships. Ships represent the most economical means to move large, heavy cargoes. Shipborne movement of goods costs less than rail transport, about one-tenth of transport by truck, and on the order of one-thirtieth of the cost by air transport. Where there is a genuine choice, such as where navigable rivers parallel roads or rails, the more economical way to move cargoes is by water transport. As Mahan wrote, "Because more facile and more copious, water traffic is for equal distances much cheaper; and, because cheaper, more useful in general. These distinctions are not accidental or temporary; they are of the nature of things, and permanent."[11]

The great seaports of the world are located in the most prosperous countries of the world, and coastal states provide virtually all of the world's catch of seafood. Littoral areas offer a large fraction of the world's most productive farmland, contain much of the world's industry, energy production, military installations, and vacation destinations.[12] Despite a few exceptions, landlocked states tend to be among the most economically backward of all. It is obvious that lack of access to the sea results in lower national economic progress.

Much of the world's general cargo moves in containers. The estimated 15 million containers in circulation make about 250 million journeys a year. Major sea lanes carrying cargo, petroleum products, and liquefied natural gas must transit narrow seas and straits in certain places. The Straits of Dover, the narrow accesses to the Baltic and Mediterranean Seas, the route around Cape Hope, the Suez and Panama Canals, the Straits of the Bosporus and Hormuz, and the narrow passages through the Indonesian archipelago—the Strait of Malacca, the Lombok and Sunda Straits—stand out as the earth's key oceanic choke points.

Choke points are important because they focus the energy of shipping, and offer opportunities for disruption of the economies of the world by terrorists, pirates, or rogue regimes. Recalling that the most difficult problem in maritime conflict is to find targets, choke points make that task considerably easier. If the ambushers know that the target must pass a certain point, that greatly relieves their "findability" problem.

Of concern in recent years has been the possibility of transporting weapons—or even people—illegally in shipping containers. Shortly after the September 11, 2001 attacks on the United States, a suspected al-Qaeda terrorist was apprehended at a port in southern Italy inside a container bound for Canada. It was reported that he had provisions for a long journey, false documents, a bed, and a bucket for a toilet.[13]

The United States has sought to address this issue by two initiatives: the Proliferation Security Initiative, and the Container Security Initiative. Aiming at the prevention of proliferation of weapons of mass destruction, the former, among other provisions, commits its participating states to "board and search any suspect vessels flying their flags in their internal waters, territorial seas, or areas beyond the territorial seas of any other state," and to "consent under the appropriate circumstances to the boarding and searching of their own flag vessels by other states and to the seizure of WMD-related cargoes." The Container Security Initiative maintains a narrower focus—specifically on containers that are being shipped to the United State—to identify containers that pose a high risk and to pre-screen them before they arrive at U.S. ports.

While the oceans are vital because of the ocean commerce they carry, they constitute an important food resource as well. Fisheries, for the most part, have been placed under the control of the coastal states by reason of being located in their Exclusive Economic Zones. Issues remain regarding migratory fish and their overexploitation, but there seems to be scant reason at present why such issues should erupt in conflict.

Maritime environmental issues will, no doubt, continue to affect the economies of seagoing states, but to date they have been effectively and extensively dealt with by the International Maritime Organization. Whether attempts to mitigate the effect of global warming will result in economic issues for maritime states, and whether these will in turn become security-related, remain as potential questions for the future.

Final Thoughts on Context

While its size and openness might imply that the oceanic realm is rather simple, composed as it is of physical, political, legal, and economic systems, the context of military operations at sea is, instead, exceedingly complex. These factors are far different from those existing on land, and have a strong conditioning effect on those who must deal with them on a daily basis.

As ships and aircraft approach the shoreline from the sea, the operational setting begins to change from an oceanic to a land environment. Clearly, the physical environment of the littoral is even more complex than that of the open ocean, and it demands that different factors be taken into account. The increasing shallowness of the water as the shoreline is approached, for example, has an effect on wave formation and height, as well as providing areas more congenial to the deployment of sea mines of all kinds and of small submarines. Reefs and shoals, tides and currents, and changes in the salinity of the water—especially in river estuaries—can affect naval operations significantly. Coves and other littoral irregularities can provide hiding places and launch areas for adversary coastal craft.

The battlespace on the ground is constrained in size by political considerations. At sea the battlespace can be global in scope, and entire formations of ships can move distances in a day that are unimaginable in land warfare. In conventional warfare ashore, the threat is generally thought of as varying in strength linearly from a center of gravity. There is no analogue at sea.

Because today anything that can be located in time and space can be targeted and destroyed, sanctuaries at sea are sought by measures to avoid detection—such as indirect routing, seeking the cover of weather, mixing in with commercial shipping, and external signature reduction—and only secondarily by active and passive defenses. To be "over the horizon" is insufficient in an era where air and space reconnaissance information can be conducted by—or even procured from—another state.

In a particularly insightful passage, Yaneer Bar-Yam wrote:

> In recent years, the military has recognized that war is a complex encounter between complex systems—systems formed of multiple interacting elements whose collective actions are difficult to infer from the actions of the individual parts. War is particularly complex when the targets are hidden, not only by features of the terrain . . . but also by the difficulty of distinguishing among friends, enemies, and bystanders. It is also complex when the enemy is divided into diverse, versatile, and independent targets; the actions that need to be taken are specific, and the difference between right and wrong actions is subtle. Complex warfare is characterized by multiple small-scale hidden enemy forces. Large-scale warfare methods fail in a complex conflict.[14]

Navies transition almost seamlessly from peace to war: the environment does not change, and operations—except for open hostilities—are not altered significantly. There are no discontinuities between peacetime and wartime operations, except (for the firing the weapons) and, of course, the accompanying risk and fear. This opens the possibility of employing maritime operations to indicate intentions or for deception, and gives rise to the use of naval forces for signaling, displays of force ("showing the flag"), and calibrated escalation control.

Taken against the contextual backdrop offered by this chapter, the war in Iraq since 2003 can be viewed as the migration over the land of many of the characteristics of contemporary naval warfare. The physical context is different, but the actions of the adversary, the non-linearity of the battlefield, and the potential for small-scale operations to have disproportionate effects are similar.

Throughout the history of naval warfare, the weaker party in an armed struggle has often resorted to irregular warfare. Commerce raiding was a key part of U.S. maritime strategy in the Revolutionary War and in the War of 1812. It was the Confederate naval strategy in the U.S. Civil War, the German strategy for its navy in World Wars I and II. Preying on commercial ships rather than warships—whether by pirates, privateers, or commerce raiders—has traditionally been the strategy of the militarily weak. When, in the late nineteenth century, Admiral Aube of the French Navy taught his *jeune ecole*, or "young school," to "shamelessly attack the weak, shamelessly fly from the strong," he provided the motto for guerrilla fighters of the twentieth century.

Guerrillas, or insurgents, do not allow lines in a battlespace to constrain them. They recognize no front, no rear, and no flanks. They can attack any time, and essentially anywhere. They do not operate on a twenty-four-hour clock, sleeping at night and eating three square meals at prescribed times every day. They know that the central difficulty for their enemy is to find them and to take away their sanctuary. One of their major counter objectives is to make their adversary afraid—all the time, everywhere. They grant no sanctuary to their enemy, and in the process have no respect for the security of innocent bystanders either.

Those accustomed to linear warfare often become disoriented when they can't find those comforting lines they had been accustomed to. Where is the forward edge of the battle area? Where is the forward line of troops? Where are the logistics lines of support? If we don't have a front, or a forward line, how can we tell who's winning? The traditional measure of effectiveness in land warfare was who controlled what pieces of real estate. That's different now in a non-linear environment. The struggle of the ground forces in Iraq to deal with an adversary who respects nothing, not even innocent human life, who welcomes death and shows little fear, has raised operational difficulties far greater even than those encountered in Vietnam. For naval forces, the struggle against Japanese suicide attacks from the air, on the surface, and underwater provided an early insight to such tactics, and the great difficulty in dealing with them.

In a non-linear context, minor actions can have outcomes out of all proportion to their military worth, which means that not only is the battlespace non-linear, but so are the potential effects. Naval officers, by virtue of the conditioning of their environment—the physical environment with no lines and great difficulty both in hiding and finding; the political environment with few rules, and a willingness on all sides to ignore even those (unrestricted submarine warfare offering the best example), and unprotected commercial activity mixed in with military forces—find much that is recognizable in the battlespaces and combat activity in Iraq and Afghanistan.

The context in which they grew up and operated at sea has greatly influenced how naval officers approach the use of naval forces to achieve national security objectives. They have demonstrated, over two centuries, that a few general guidelines have oriented their thinking and resulted in success. The consistency, and the persistence of these guidelines have been remarkable over the centuries. The following chapters will reveal them, how they arose, and their potential for informing naval strategy for the future.

CHAPTER 4

Strategies for the Employment of Naval Forces

Strategies describe the *ways* organizations apply *means* to accomplish their *ends*, or objectives. Military strategy is about choices of how to apply military force among a variety of ways to achieve military, leading to the fulfillment of political, objectives. It is about the *context* in which means will be applied, and about anticipation. Strategy seeks to foresee what a thinking, motivated, ruthless, clever, vicious adversary might do, and then to craft ways to thwart it. It attempts to devise ways to attain the greatest leverage in using force decisively while causing the least collateral damage. Strategy strives to strip opponents of their sanctuary while preserving one's own.

A selected course of action invariably involves mismatches among ends, ways, and means, the measure of which is operational risk. To put it slightly differently, if the available means (number or type of weapons, for example) are deemed insufficient to execute a chosen strategy, then there is a tangible risk of failure. One must then depend for success on luck, enemy mistakes or, perhaps, guile. High risks obtain if a strategy appears incapable of achieving given objectives.

Only rarely is there a precise correlation between ends, ways, and means. The normal situation is that means are limited, and struggles arise over choices either of ways to apply them or of objectives to be sought. Moreover, ways, means, and ends are in constant interaction. Feedback mechanisms continuously interact with and affect the choice of means and the ways they are applied. Success or failure of some ways can result in the modification of ends or, alternatively, the application of different means. Technical or operational surprise or setbacks in execution (by-products of actions taken by the adversary); the failure of weapons to perform up to design standards; or the negative consequences of poor training—will significantly alter one's approach. Incomplete or inaccurate intelligence is a constant concern in any conflict, and it manifests itself in increased levels of risk. The interplay among such key factors in conflict is dynamic and unrelenting so long as the struggle continues. As a consequence of these simultaneously moving parts, risk will inevitably accompany the employment of military force. The risk is to mission accomplishment, which might result in loss of the lives of one's citizens and, ultimately in an existential war, loss of freedom.

Given the uniqueness of the context in which conflict takes place in maritime areas and the cultural predispositions of the practitioners of the employment of armed force there, it would stand to reason that strategic approaches for the employment of military forces would be starkly different on sea than on land. And so they are.

This is hardly a dazzling original insight. Nearly half a century ago Bernard Brodie wrote: "Naval warfare differs from land warfare in the objectives aimed at, the implements used, and the characteristics of the domain on which it is waged. These are major determinants indeed, and we must therefore expect maritime strategy to differ materially from the strategy of land warfare. Certain ideas are common to both, but it is a mistake to carry over wholesale into the realm of the sea, as some writers have attempted to do, concepts which govern the conduct of war on land."[1]

Yet, the study of strategy as an intellectual discipline extends back only until the nineteenth century. Surely, Thucydides and Sun Tzu wrote books that offered strategic insights for military thinkers, and Napoleon contributed battlefield wisdom to the publication of his *Maxims*. National, or military strategy, however, became the focus of directed study only with the works of Jomini and Clausewitz and later in the nineteenth and early twentieth centuries—for naval warfare as a distinctly different form—by Mahan, Colomb, and Corbett.[2]

Warfare in the maritime arena tends to be protracted, and for a host of reasons, including: the global size of the battlespace and its potential to provide sanctuary, the relatively slow movements of waterborne platforms on and under the surface of the oceans, the absence of geographic or functional lines to organize and focus action, and the ability for oceanic military forces to avoid conflict. In addition, the decisiveness and irrevocability of the losses of capital ships, coupled with their low overall numbers, militates against employing them recklessly. Indeed, the great British strategist B. H. Liddell-Hart noted that combat at sea did not allow blitzkrieg tactics; he likened sea power to the slow-acting effects of radium: "beneficial to those who use it and are shielded, it destroys the tissues of those who are exposed to it."[3]

In view of these considerations, states have historically adopted six discrete national strategies for the employment of their maritime forces: three offensively oriented (fleet battle, blockade, and maritime power projection) and three defensive in nature (commerce raiding, fleet-in-being, and coastal defense).

The first three strategies—fleet battle, blockade, and maritime power projection—constituted strategies typically adopted by states endowed with large, powerful navies. Broadly these can be described as offensive strategies, for they seek by sustained positive action to win battles, campaigns, and wars. The latter three—commerce raiding, fleet-in-being, and coastal defense—were embraced by those with more unexceptional maritime military means. These are considered to be essentially defensive strategies, for their goal is focused on avoiding defeat rather than securing victory.

All of the strategies pivot on the central problem of naval warfare: finding the adversary. Surveillance and counter-surveillance—finding and hiding—constitute the first order of business. The lone exception to this general rule is the fleet-in-being strategy, for the fleet-in-being wants its presence known and felt, but it does not seek battle.

Offensive Strategies for the Employment of Navies

At the heart of all offensive strategies is the concept of "reach." "Reach" encompasses the capability to concentrate and employ military forces at long distances to decisive effect. To strike a single target or five targets at intercontinental distances with long-range missiles does not constitute "reach." It is a function, instead, of the placement of offensively capable forces in areas non-adjacent to the homeland, to be able to conduct operations from those forward positions, and to exploit the advantages gained. The sustainable operational range and striking power of weapon systems, robust logistic support, and well-functioning command and control all fill important roles in underwriting reach. Clearly, bases in the vicinity of areas of operations are very useful in attaining and leveraging reach. Equally clear is the fact that few countries in the history of the world have enjoyed a high level of strategic or operational reach. Furthermore, it does not go unnoticed that such capability—to be able to project military force on a wide geographic basis—bestows substantial political leverage on its owner.

Fleet Battle, a strategy proclaimed by Mahan to be the preferred method of gaining the unhindered ability to use the seas, meant maintaining a strong battle fleet, and employing it to drive the adversary fleet from the oceans. The most recent practitioner of Fleet Battle as a core naval strategy was the Imperial Japanese Navy.

Before World War II, Japanese strategists attempted to divine how large and powerful a battleship the U.S. Navy might build. "Any U.S. battleship," they reasoned, "would have to be able to transit the Panama Canal. Mahan's exhortations not to divide the fleet will powerfully influence U.S. battleship designers."[4] The Japanese Navy calculation was that the width of the canal would determine the maximum beam of U.S. battleships and, therefore, based on sound architectural principles for design of warships, their length, displacement, and maximum size of embarked guns. Intuiting U.S. shipbuilding proclivities in this way led to the design and construction of IJN *Musashi* and IJN *Yamato*, the largest and most heavily armed, and armored, battleships in history. Given these inputs, they determined that the U.S. would build a ship about the size of the *Iowa* class. *Musashi* and *Yamato,* laid down in 1937, three years before the first of the *Iowas,* carried 18.1-inch guns to the comparatively inferior 16-inch guns of the *Iowas.* Yet, a fleet engagement between them never occurred: all but one key battle in the Pacific were profoundly influenced by airpower—before the largest battleships came within gunfire range of one another. The exception was the Battle of Surigao Strait, in which the first-line battleships of neither nation participated.

In modern interpretation, fleet battle could incorporate within its definition such actions as offensive naval operations against adversary submarines—especially ballistic missile launching submarines. Superior, aggressive navies—such as the Royal Navy under Nelson, or Halsey's U.S. Third Fleet in World War II—seek fleet action to attain maritime superiority.

Maritime Blockade, controlling the movement of adversary ships or prevention of shipment of contraband cargoes in either adversary or neutral ships, also requires superior naval forces.[5] Depending on the number of ports of shipment involved and the geographic separation between them, the inventory of blockading forces required to mount a successful blockade can be very large. British maritime blockades of France and its coalition members in the seventeenth and eighteenth centuries—since they were required simultaneously to guard ports on the English Channel, in the Atlantic, and in the Mediterranean— were brutal in terms of their impact on Royal Navy ships and sailors. Far more British sailors died on blockade station from sickness and disease than perished in battle against the French or combined French-Spanish fleets.

Historically, blockades have been difficult and expensive to undertake. More over, blockades tend to be penetrable. The incidences of ships (and cargoes) penetrating blockades are frequent in the history of naval warfare. Before the invention of submarines, sea mines, and long-range artillery, blockades were "tactical" or "close," requiring blockading ships to sail near to enemy ports and maintain a visual watch on blockaded ships. At night, on a favorable wind, or in a fog, even such a tight tactical blockade could be breached. Subsequently, in the age of "strategic" blockade, ships were obliged to keep greater distance, and breaking out of the blockade became easier.

Blockades can exert great stress on populations. Perhaps three-quarters of a million civilians died as a direct result of the Allied naval blockade of Germany during World War I. By comparison, fewer than two million people have been killed by aerial bombing in all wars combined.[6] Blockades are prolonged actions, often extending for years, while battles in comparison—even the greatest sea battles of all time—have lasted only a matter of hours. In recent times, since opposing naval powers have declined significantly in number, blockade has come to be focused more on methods to prevent the waterborne movement of specified cargoes—in particular, the shipment of technologies involving nuclear power, long-range offensive missile systems, and weapons of mass destruction.

Mahan's insight on blockade is wholly relevant today. Over a hundred years ago he wrote: "Whatever the number of ships needed to watch those in an enemy's port, they are fewer by far than those that will be required to protect the scattered interests imperiled by an enemy's escape. Whatever the difficulty of compelling the enemy to fight near the port, it is less than that of finding him and bringing him to action when he has got far away."[7] The U.S. Proliferation Security Initiative and Container Security Initiative operate on this principle by seeking to control the shipment of specified cargoes principally at the point of origination of shipping.

Maritime Power Projection means, straightforwardly, the employment of military force against landward targets from maritime positions. It includes amphibious operations, carrier and missile strikes, special operations, and other offensive operations to support such efforts, an example of which is mine-clearing operations adjacent to a hostile coastline.

Historically, naval forces have had difficulty in directly attacking enemy centers of gravity or, in other words, coming to grips with the sources of enemy power, which are invariably located on land. Accordingly, as noted earlier, strategies for the employment of naval forces have typically taken extended periods of time to exert their effects. Decisive battles among fleets, although brief once action is joined, have been few and far between. Moreover, their impacts have often taken years to be realized.

Blockades tend to be notoriously slow-acting. Power projection, therefore, as the most direct expression of naval power, and as a way to achieve strategic objectives comparatively quickly and decisively, has come to be emphasized in recent times. In particular, when nuclear weapons went to sea in ships an additional dimension was added to power projection.

Before the advent of long-range bombers or ballistic missiles with ranges in the thousands of miles, it was evident that nuclear weapons would have to be deployed relatively close to their targets. For the United States, that meant positioning nuclear weapons using bomber aircraft and missiles in the European and Western Pacific theaters to confront the Soviet Union, and for the Soviet Union, the provision of short-range aircraft and missiles to oppose nearby adversaries.

The Soviet Union, separated by the oceans from its main enemy, the United States, exploded its first atomic weapon in 1949, four years after the initial U.S. atomic test. Having no allies in close proximity to the continental United States, the Kremlin felt compelled to develop not only long-range ballistic missiles—the shockwave projected by the launch of *Sputnik* in 1957—but also to place them on submarines.

Submarines were attractive because the freedom to use the open seas and the sanctuary they provided enabled the ships to be stationed close to their targets on the North American landmass. Consequently, the Soviet Navy launched the world's first ballistic-missile carrying submarine with the diesel-powered *Zulu* class in 1956. The *Golf* class followed quickly, and thereafter the Soviet Union produced four different classes of nuclear-powered submarines capable of launching ballistic missiles, the *Hotel*, *Yankee*, *Delta*, and *Typhoon* classes.

The United States took nuclear weapons to sea in its aircraft carriers in 1951, and on cruise missiles in aircraft carriers, cruisers, and submarines with the *Regulus* program in 1956. The five diesel-electric submarines that participated in *Regulus* were obliged to surface in order to fire their missiles, so submarine deterrent patrols that began in 1959 in the Pacific Ocean were superseded quickly by nuclear-powered submarines carrying *Polaris* ballistic missiles in 1964. U.S. ballistic missile submarine patrols in the Atlantic and Mediterranean areas had begun in 1960.

Long-range ballistic missiles were launched from geographically fixed positions, which made them theoretically, at least, vulnerable to a pre-emptive attack. Mobile, stealthy submarines were by comparison very secure from attack; and thus, as nuclear missile launching platforms, could act as a strategic reserve force. This was the role that both the Soviet Union and the United States assigned to its ballistic missile launching submarine fleet.

In nuclear strategy, it is vital to maintain a secure strategic reserve force.[8] A strategic reserve is a force that can survive an adversary's first strike and, in retaliation, render unacceptable damage to the aggressor. It was called, during the eras in which Mutual Assured Destruction was in effect between the United States and the Soviet Union, the "secure retaliatory force." In long-standing, clearly understood, military terms, however, it was a strategic reserve.

The most important characteristic for a strategic reserve force is that it be secure: it must not be able to be attacked credibly by the adversary, for it is the force that either secures victory in war or reliably avoids defeat. Second, it must be of sufficient size to make a strategic difference if it is called upon. And finally, it must be controllable: one must be able to communicate with the force and control it, even after having suffered a devastating nuclear attack.

A paradox of nuclear strategy is that one can use a strategic reserve only in the most dire circumstances, for if one expends it and it is ineffective, victory goes to the other side. This means that unless success can be guaranteed, the possessor of a strategic reserve force would never use it. Clearly, success can rarely, if ever, be guaranteed, so there are powerful reasons for never actually using a strategic reserve. Its value is in having it, not using it. "Use it or lose it," under this reasoning, is a question that those responsible for a state's security, and not prone to a suicidal spasm of irrationality, would likely never face with respect to their strategic reserve force.

One of the key aspects of nuclear strategy is to find ways to nullify the other party's strategic reserve. If the adversary does not possess a strategic reserve, or maintains a strategic reserve that is vulnerable, it runs high risks of defeat in war.

Strategic nuclear reserve forces can be provided in a variety of ways. For example, having a strategic nuclear force resident in three separate deployment modes, greatly complicating an adversary's ability to destroy it in detail, is one method. The United States had, for many years, a strategic nuclear policy of maintaining a strategic "triad" composed of intercontinental-range bombers, land-based ballistic missiles, and ballistic missile-launching submarines. No adversary could, with confidence, attack all three legs of the Triad simultaneously with confidence that it would not suffer a devastating counterattack from the remaining "reserve."

Within each leg of the triad, security was provided in a variety of ways. Bombers were initially placed on airborne alert. Then they were moved away from the coastlines so that they could not be attacked by missile-launching submarines, then they were put on ground alert status. An airborne national command post was in use for many years so that command and control of strategic forces could be assured during, and even after, an attack.

The land-based missile force underwent efforts to harden and disperse it so that it could survive all but the most accurate, powerful strikes. (The more accurate an attack the less powerful it needs to be to destroy the target.) Consideration was given to making the land-based force mobile (the Mobile Protective Shelter and rail mobility programs), placing the missiles deep underground, defending them with ballistic missile defenses (the Safeguard program), and even placing the missiles so close together that some would have an increased probability of surviving an attack (called "Dense Pack"). For bombers and missiles, since they can be located without great difficulty, their effectiveness had to be underwritten by defending them, hardening them against the effects of an attack, dispersing them, or by making them mobile.

For submarines, which by virtue of their underwater deployment are orders of magnitude more difficult to find than bombers or land-based missiles, the problem of their survivability was approached differently by the United States and the Soviet Union. The United States for many decades enjoyed technological superiority over any potential adversary in its submarine nuclear power plants, in submarine quieting (reducing the submarine's sonic signature so that others could not detect it), and in submarine-launched ballistic missile range and accuracy. As operational missile ranges increased through successive generations of missiles, the United States adapted by increasing the size of the patrol areas of its submarines. Along with quieting technologies and deeper diving capabilities, the increase in patrol area size, while still covering the same or even larger target set in the Soviet homeland, resulted in an increase in sanctuary for U.S. ballistic missile submarines.

The Soviet Union went about providing for the security of its submarines in a markedly different way. While it also worked on improving the reliability and capability of its nuclear power plants and the quietness of it submarines, it used the extended missile ranges it developed over time to move its submarines into protected locales near the Soviet homeland. In those close-in areas, called "bastions" in the West, the Soviet Navy could provide three-dimensional protection for its ballistic missile-launching submarines—by aircraft patrols, by anti-submarine capable surface ships, and by quiet attack submarines. By having its strategic reserve force nearby the Soviet homeland, moreover, the Soviet Navy argued that more reliable command and control could be maintained over it, even in the midst of a nuclear war. This was the Soviet approach to providing the greatly needed and valued safe haven for its strategic reserve, its ballistic missile-launching submarines.

The contrast between the two approaches is worthy of note. The United States relied on the large, featureless open ocean areas and depths, coupled with superior platform quietness, to offer sanctuary for its stealthy submarine force. The Soviet

Union, more concerned about the survivability and controllability of its ballistic missile-launching submarine force, was unable to use the leverage available to it of the open ocean, and obliged to devote additional military assets to its protection.

The contributions that naval forces make to the overall military strategies of the states they serve have value only insofar as they can influence political processes, which invariably take place on land. To sink an enemy fleet in isolation from events taking place ashore is to have accomplished nothing. Blockades that fail to alter policy are impotent. Power projection that does not succeed in deflecting the actions or intentions of an adversary is squandered.

In contrast to these three offensive strategic options, strategies that embrace a more defensive orientation are generally undertaken by states by default, in the main because a paucity of naval forces results in an inability to pursue any of the offensive approaches.

Defensive Strategies for the Employment of Navies

Commerce Raiding, attacking the adversary's merchant shipping, has a long history, extending back to the first days of organized oceanic commerce. A strategy of commerce raiding, such as was conducted by the United States in the Revolutionary War and the War of 1812 against Britain and most infamously in both World War I and World War II by Germany, is differentiated from, say, piracy, in that it is conducted by a sovereign state or other organization for the purpose of achieving military or political objectives. Piracy, on the other hand, constitutes illegal actions conducted on or over international waters by private individuals for private ends, usually monetary gain. Piracy is prohibited by U.S. and international law, while commerce raiding is the subject of regulation, but is not proscribed by the Laws of Armed Conflict.

Using small ships, and later submarines, as commerce raiders was popular in states with small navies such as France and Russia, who viewed such a strategy as an asymmetrical way to gain leverage from a small navy, and by the Germans in both world wars as a method to disrupt the logistic support of the war by sea. As noted in Chapter 3, it provided an early intellectual backdrop for guerrilla warfare.

Commerce raiders employ speed and stealth to avoid action by opposing naval forces, focusing on less well-defended merchant ships. Unrestricted submarine warfare by Germany brought the United States into World War I. In an ironic twist, the United States declared unrestricted submarine warfare on Japan immediately after the attack on Pearl Harbor. The use of submarines in this fashion against

Japan constituted a hybrid strategy that combined defensive and offensive forms of naval warfare, commerce raiding and blockade.

The term "fleet-in-being" is not well understood by other than naval strategists. It refers to maintaining a fleet that is just strong enough that others, however superior they might be, will not attack because the risks of failure appear too great.

The classic example of the fleet-in-being strategy is the pre–World War I "risk fleet" strategy of German Adm. Alfred von Tirpitz. Acknowledging that the German High Seas Fleet would likely not ever be able to confront Grand Fleet without disastrous results, Tirpitz opined that it would be sufficient for Germany to maintain a fleet that was just strong enough that the British would not seek battle with it out of fear of losing (given the fact that Britain had other adversaries who would benefit by a weakening of its naval power, as well as a far-flung empire to protect). The only major naval battle of World War I, fought off Jutland, was an attempt by the Germans to lure the Grand Fleet into minefields off the coast of Denmark, not to engage the Grand Fleet.

As noted previously, the fleet-in-being wants its location and composition known to the adversary, but believes that the force is too strong to be attacked successfully. Late in World War II the German battleship *Tirpitz* was moved to a Norwegian fjord, where, as a single-ship fleet-in-being, it exerted a powerful influence on British and Allied strategy. The intention was to encourage the Allies to position forces in the North Sea to guard against a sortie by *Tirpitz* that would endanger the convoy routes to Murmansk. Forces the allies were obliged to keep on watch for the *Tirpitz* were unavailable for operations elsewhere. The story of the herculean efforts expended by the British to pull this strategic thorn in the Royal Navy's paw reached culmination in the employment of twelve-thousand-pound "Tall Boy" bombs on *Tirpitz*, dropped by thirty-one Royal Air Force *Lancaster* bombers.

In the end, Tirpitz's World War I strategy failed because it hinged on an assessment that Britain would not attack and thereby make itself more vulnerable to other potential adversaries, namely, France and Russia. As it happens, France and Russia allied with Britain against Germany and the "Risk Strategy" of Tirpitz was ineffective. Certainly, the inconclusive action at Jutland, however, indicated that it was nearly successful, for Winston Churchill remarked: "Jellicoe [the Royal Navy commander at Jutland] was the only man on either side who could lose the war in an afternoon."[9] The World War II obliteration of the battleship *Tirpitz*, on the other hand, demonstrated that a determined adversary can destroy a fleet-in-being if it is willing to devote enough assets and energy to the effort.

The third defensive option for maritime strategies is *coastal defense*. States with a coastline are obviously vulnerable to attack from the sea, and in the twenty-first century states have become more and more sensitive to the security of their seaward frontiers. In 1959, for example, only sixty-seven countries supported national navies, but in less than two decades, by 1977, that number had doubled. Not counting those states that maintain maritime forces solely for police, customs, or coast guard roles, in 2008 the number of organized navies in the world is about 110.

While the term "coastal defense" seems self-explanatory, it constitutes an amalgam between functions undertaken in wartime against an adversary with seagoing forces, and policing the areas near the homeland against drug smuggling, illegal immigration, infiltration of enemy agents across the coasts, piracy, slavery, environmental damage, protection of fisheries, trade interdiction, weapons proliferation, and terrorism. Most states of the world are relegated to this approach to maritime security simply because their navies have neither the numbers nor the endurance—either in fuel capacity or in replenishment capability—to operate far from home waters.

States choose among these six strategies in accordance with their objectives and the means available to accomplish them. Moreover, as often as not, the strategies are not pursued separately but combined or pursued sequentially—sometimes even concurrently—during the course of a conflict. For example, in the Pacific Ocean areas in World War II the United States undertook fleet engagements—most notably at Midway—and conducted power projection campaigns across the string of Pacific islands, while blockading Japan with submarines.

The United States today has no adversary fleets to engage, nor may it reasonably expect to for the time being. Commerce raiding is incompatible with achieving U.S. objectives. Fleet-in-being strategies have historically been used by weak navies for purposes of deterrence or defensive response. For more than a century the United States has been the preeminent practitioner of "forward presence"—employing naval forces away from its homeland to deter adversaries, to reassure allies and friends, and to shorten the time for crisis response. Unquestionably, forward presence importantly complements a strategy of maritime power projection.

Increased participation in the homeland defense mission will involve the employment of ships as a sea-based adjunct to national missile defense, and also to extend missile defense umbrellas over the territories of friends and allies. In addition, the future security environment promises to require an increase in U.S. coastal surveillance and reconnaissance, and in patrol capability. Anti-smuggling, anti-infiltration, and ship inspection functions at the more than 350 American ports taxes current and programmed U.S. Coast Guard assets significantly. Unquestionably, the number of units assigned to these tasks will increase, and the extra burden will have to be shouldered by the Navy—given the Coast Guard's size and breadth of assigned duties, which include major devotion to at-sea public safety and rescue, interdiction of maritime drug trafficking, and protection of American fisheries. The increased requirements are for air reconnaissance and surveillance, and for numbers of small ships, minimally armed. One advantage enjoyed by the United States that, in general, is not shared among its allies, is that most U.S. ports (except those close to Mexico or the Caribbean islands) can be approached only by capable, seagoing vessels. The threat of infiltration or smuggling by means of submarines, while it cannot be ruled out entirely, appears unlikely and small enough not to devote tailored resources to it.

Note the positive relationship between the prospective effectiveness of a strategy and the degree to which the adversary's sanctuary is threatened. Recent emphasis on power projection and the change in the nature of maritime blockades can also be viewed as a by-product of the atrophy in the size of opposing naval fleets. The lack of opposition to their establishment of sea control has permitted the few large and powerful navies to reorient their focuses in a landward direction. It has also given rise to studies of "anti-access" capabilities of potential U.S. adversaries—the use of "asymmetrical," often non-naval capabilities to thwart U.S. expeditionary power projection. As an effort to counter the Strategic Culture, anti-access approaches will be examined in Chapter 8.

Sea Control and the Sea Lanes

One might remark that a core competency of effective navies—control of the seas—has nowhere been mentioned in the discussion of strategy options. Control of the seas, and in many instances of the air and space as well, is, nonetheless, embedded within each of the offensive strategies. Control of the seas means that one can use the seas for one's own purposes, and one's adversaries cannot. Control of the seas is a local phenomenon, regarded to be limited in space and time. It is also understood not to be static and comprehensive, but to be dynamic and contested. Control of the seas, consequently, is best described as a mission or function.

There is no *strategy* to control the seas; instead, control of the seas is a mission within a strategy. Control of the seas is differentiated from, but frequently misunderstood for the term "command of the seas." The latter phrase is properly used to mean that adversaries are incapable of interfering with the effective operational use of the seas, or of sea commerce, anywhere at any time. In military planning one seeks control of the seas, because that is linked to specific operational tasks in the execution of strategy. Command of the seas is a consequence of rendering adversaries incapable even of effectively contesting control of the seas anywhere.

Defensive strategies do not seek sea control because they are, for the most part, adopted out of a position of weakness. Defensive strategies attempt to prevent the adversary from establishing sea control, which in most cases is either prerequisite to or the by-product of offensive strategic options. Effectively challenging attempts at gaining sea control is called "sea denial." This is a condition in which one cannot use the seas for one's own purposes, but the adversary is prevented from doing so as well.

Another hallmark of strong navies is "protection of the sea lanes," yet that notion appears nowhere in the six strategies either. This is because the term "sea lanes" is understood by naval strategists to be a concept, not a place. "Sea lanes" are where ships travel across the oceans. Unlike roads on land that exist regardless of whether trucks, cars, or buses travel on them, sea lanes do not exist if ships are not present. The exception is where there is only one, or very few, routes to transit from point to point. Entrances to ports typify areas in which there are geographic sea lanes, and, of course choke points gain their prominence from this fact as well.

Port entrances and choke points are key to maritime strategies because they are important aids in solving the central problem of maritime warfare: finding the adversary. In fact, as previously noted, most sea battles throughout history have taken place within the sight of land because ships on the open ocean are difficult both to find and to bring to battle. Still, if there are no ships present, no protection of the routes into ports or through straits is needed. So, protection of the sea lanes amounts to the protection of ships, not geographic lines or areas. Offensive and defensive missions are embedded in all of the strategies, whether the mission is establishing and maintaining sea control or protecting ships as they travel across the seas.

Protecting the sea lanes can be accomplished in the vicinity of the protected force, or it can be undertaken at long range. For example, blockading enemy submarines in their home ports is one method of ensuring the safety of friendly ships plying the sea lanes. This was the purpose of the offensive minefields laid in both World Wars by the allies. Mines alter geography. Accordingly, offensive minefields are used to deny the enemy the use of certain sea areas, while defensive minefields are laid in one's own waters to obstruct attack from the sea. Convoying was demonstrated in the world wars as a method of reducing attrition of transiting ships.

The sea lanes can be attacked either by attempting to sink the ships at sea, or they can be attacked at their terminals. Ports of embarkation and debarkation offer a limited number of *fixed* targets to attack, in contrast to ships—a considerably larger number of *moving* targets. The Soviet Union's naval strategists demonstrated their understanding of this point, for in their publications they were careful to point out that the number of ports from which (ports of embarkation) and into which (ports of debarkation) reinforcement and resupply shipping could be routed in time of war was small, and constituted a priority target. Thus, in the Soviet view, massive flows of shipping from the United States to a European theater in wartime, as occurred in the two World Wars, would more effectively be interdicted at the shipping terminals rather than at sea.

From this brief, compressed discussion of naval strategy, it is a simple matter to deduce that the United States, since World War II the world's preeminent naval power, has shunned the defensive strategic approaches to maritime warfare in favor of those with offensive purpose. Central to Navy Strategic Culture has been the seminal idea of "attack at the source." Destroying adversary forces at their berths in port, at their airdromes, in their submarine pens, or in their garrisons was preferred by far to finding them in the battlespace once they had dispersed.

Pivotal to the U.S. approach also has been the propensity to favor expeditionary operations—forward and highly mobile. Adaptability, grounded in large degrees of self-reliance and underwritten by flexible forces, has been another of the hallmarks typifying naval strategic thought and action. All of this is bound together by a systems approach to operations and ultimately to strategy, a deep understanding of the central importance of technology, and an innate integration of all of these matters into assessment and acceptance of risk. Chapters 5 and 6 will turn to these subjects—ones that through a familiarity of history, the development of common understandings, and the sharing of mutual experiences have become the precepts of Naval Strategic Culture.

CHAPTER 5

Expeditionary

The place was the Maghreb—the Arabic "land of sunset" on the Mediterranean coast of North Africa. It was early in the life of the American Republic, yet many of the recurring elements of U.S. history were already in place. U.S. national interests had been unlawfully attacked; the free flow of international commerce had been challenged; Americans had been terrorized, taken hostage, enslaved, and abused. The Muslim "Barbary" states (Algeria, Morocco, Tunisia, and Libya) had taken advantage of the geographic choke point into the western Mediterranean at Gibraltar to prey on shipping—capturing crews and passengers and stealing cargoes. For a century and a half, from the middle of the sixteenth to the end of the eighteenth centuries, over a million Europeans and Americans were captured. In fact, more slaves were brought from Europe to the Muslim states of North Africa than were shipped from all of Africa to the United States.[1]

Having freed itself from British colonization, the United States had, as a consequence, lifted the mantle of protection of the Royal Navy from American shipping in the Mediterranean. The ensuing policy debate within the fledgling former colony about how to deal with the now-exposed American shipping was fierce, even without the amplification of an instantaneous mass media. On the one hand there were those—notably John Adams—who would, as a matter of policy, pay tribute in an attempt to appease the pirates. Taking a long view, Adams argued that "we ought not to fight them at all unless we determine to fight them forever."[2]

Nevertheless, Thomas Jefferson was president, and as the fraction of national revenue dedicated to tribute topped 20 percent, he sent Navy ships to establish a presence in the western Mediterranean to protect ocean commerce, enforce treaties, and terminate the outflow of public funds paid to the tyrants in tribute. The historical parallels to the present time are unmistakable. When Jefferson and Adams traveled to London to meet with the Tripolitan envoy in 1786 they were told: "It was written in their Koran that all nations who should not have acknowledged their authority were sinners, and that it was their right and duty to make war upon them wherever they could be found, and to make slaves of all they could take as prisoners."[3] The echoes in Osama bin Laden's fatwa of 1998 are palpable, but striking nevertheless: "We . . . call on every Muslim who believes in God and wishes to be rewarded to comply with God's order to kill the Americans and plunder their money wherever and whenever they find it."[4]

In historical perspective, the sinews of Navy Strategic Culture were forged in offensive operations "from the sea" against the Barbary pirates: naval blockade, shore bombardment, special operations, and amphibious assault. Their colorful strands—blue, green, brown, and red—can be traced down through the decades to the present time. Blue threads stretched continually through the vast open reaches of the sea, reflected in the uniforms of sailors; green in the littoral areas in which the major battles and support operations have taken place in naval warfare, expressed in the uniforms of Marines; brown in the Marine Corps campaigns ashore (and Navy riverine actions borne with little relish); and red in the blood shed to secure national political objectives, underwritten by military action and emblematic in the red stripe along the outer side of Marine Corps dress blue uniform trousers.

The Marine Corps has long enshrined the memory of those campaigns of the early nineteenth century, most notably in the "Marine Corps Hymn": "From the Halls of Montezuma to the Shores of Tripoli. . . ." The continuity with U.S. Navy operations, and the historical contribution to naval culture, have only recently been fully realized.

From the North Africa campaigns, which underscored the fact that the United States had an enduring stake in global affairs, the better part of a century would pass before Alfred Thayer Mahan penned his historical treatises on naval warfare. During that century, the fortunes of the Navy waxed and waned, from the nadir of Jefferson's "Gunboat Navy" through the Mexican War and Civil War to the zenith of the Spanish-American War. The progression of conflicts, technological change wrought in the Industrial Revolution, and theorizing on naval warfare—not only by Mahan, but by other seminal thinkers on the subject such as the Colomb brothers, Julian Corbett, and those present at the founding of the Naval

War College at Newport, Rhode Island in 1884—all contributed to shaping the Navy Strategic Culture.

From the earliest part of the nineteenth century until today, U.S. naval forces have been deployed in forward areas around the world.[5] Then, as now, the Navy was ensuring that the seas of the world would be fully open and free for the commercial use of all; and, by its global presence claiming a place at the table in deciding what changes were to be made in the political landscapes and seascapes of the world. It has been deployed forward and ready, fully ready, at all times to react to crises with military action and to emergent humanitarian needs with aid.

The expeditionary nature of the Navy–Marine Corps team was forged on the shores of North Africa over two hundred years ago. Naval forces were to be forward-deployed, mobile, offensive, and adaptable. Navy Strategic Culture has been shaped by two major themes, one contextual, and one action-based. The former—comprising the physical, political, legal, and economic environment—was detailed in Chapter 3. Here the second theme that conditions Navy Strategic Culture—the action theme—will be detailed.

The key word is *expeditionary*. The resonance the term incites cannot be denied. Going afar to pursue the objectives of the state by establishing presence and projecting power: that is what it is all about. The keys to being expeditionary are to be forward, mobile, offensive, self-reliant, and adaptable.

Forward

From the start of the republic, and from its own inception, the U.S. Navy sought to take advantage of the great maritime commons of the seven seas by sailing actively and globally. U.S. commerce needed a shield; pirates and slavery, to be suppressed; scientific expeditions and surveys, to be pursued; whalers, a cover for their exposure in the great reaches of the Pacific Ocean; the American flag, representation; victims of natural disasters, relief; and U.S. envoys, protection in capitals all around the world. U.S. Navy ships have long positioned themselves forward to provide a tangible measure of U.S. intentions. They pay no political price for these operations, for the high seas constitute the freeways of sea power, and ultimately of national power. Throughout history, the only other Navy with such pretensions of global responsibility and effort was Great Britain's Royal Navy, which was further burdened by the requirements of maintaining a far-flung empire. "It is not so much actually in order to shoot her guns at England that America wants her cruisers; nor to defend her supplies from American cruisers that England is so particular about her superiority at sea. It is because naval preeminence means international prestige; preponderance in the counsels of the world; authority in troubled areas such as China; power to have one's way; political backing to financial economic and commercial penetration," wrote Salvador de Madariaga.[6]

Peacetime presence is founded on, and fueled by, the premise that the active employment of non-violent power, often referred to as "soft power," is the best way to reduce the necessity for a subsequent use of force. That is, the willingness to expend time, effort, and funds to be present, on the scene rather than stationary in far distant garrison, conveys a strong interest in what is taking place, and both willingness and readiness to become involved, if necessary. This forward presence helps to deter adversaries, to influence potential adversaries, and to reassure friends and allies of U.S. commitment.

Forward presence means that deployed forces, over time, become familiar with local conditions and environments, facilitating intelligence gathering, reconnaissance, and surveillance. It ensures that reaction times to make available formidable striking power will be short. Being routinely forward and anticipating where problems might arise helps make certain that U.S. forces will be the first on the scene. Because the high seas are free for all to use, moreover, it is axiomatic that the second party to arrive at the scene of a crisis, not the first, increases the risk of conflict in the maritime arena.[7] As a consequence, the forward placement and ready configuration of naval forces leaves to the adversary the decision to break the peace. Without forward basing, Army and Air Force units cannot be employed in this way. The involvement of Army or Air Force assets generally requires that hostilities have already commenced—because the sovereignty of another must be infringed upon—or that the preemptive use of force has been decided upon. While relocation of such forces might be viewed as an indication of intent, it tends to reduce flexibility and constrain options.

Homeland defense, from the outset, was viewed by naval officers as a garrison operation, more suitable to the disposition and capabilities of the Army. Once U.S. national borders had been stabilized, and the threat from without minimized, interest increased in expanding national horizons. The initial manifestation of that new "destiny," once consolidation of the lower forty-eight states was complete, was in the writings of Mahan, made concrete by the Spanish-American War.

U.S. ships had been cruising the oceans for nearly a century before the Spanish-American War, sometimes more frequently and farther afield, sometimes less. The Navy was throughout the nineteenth century, as it would be for the first half of the twentieth, primarily captive to the political, and by extension, the economic policies of the administration in power in the government.

Homeland defense, with a few expeditions interspersed, was the order of the day before the War of 1812. For the remainder of the century, with the exception of the period of the Civil War (1861–1865), the forward operations of the navy ebbed and flowed, but were never entirely absent. Squadrons of ships occupied stations in the Indian and Pacific Oceans, in the Mediterranean Sea, and in the North and South Atlantic. During the Spanish-American War battles were fought in the Caribbean, and in the Philippine archipelago; and not long after that war was concluded, President Teddy Roosevelt sailed the Great White Fleet of battleships on the world cruise of 1907–09. During the Mexican War of 1846–48 and the U.S. Civil War, U.S. warships took up forward stations to restrict the maritime movements of their adversaries, and to project power ashore.

In the Spanish-American War the Navy fought fleet battles off Cuba and in Manila Bay. These were conducted in the manner of Mahanian fleet-on-fleet operations, and they were successful in defeating the Spanish Navy and contributing strongly to winning the war. Then, in World War I the Navy operated in the European theater, and was instrumental in safeguarding the transport of war-supporting cargoes and two million troops to the front. All U.S. ground forces traveled by ship to Europe, and although the allies lost over five thousand ships during the war, no troopship was sunk on eastward passage. World War II found the Navy in forward Atlantic and Pacific operations, and escorting convoys in the former theater. This occurred despite Japanese incursions on the U.S. West Coast and German sinkings of merchant ships within sight of East Coast cities. The exercise of airpower over Korea and Vietnam, along with blockade and riverine operations in the southern part of the latter country, typified naval action in those two wars.

Permanent forward presence has been epitomized by carrier battle groups in the Mediterranean Sea since 1947 and in the western Pacific Ocean since 1950. Naval forces have maintained continuous presence in the Persian Gulf, although in a lesser force posture than other areas for the most part, since 1949. At present, expeditionary U.S. naval forces comprised of many types of ships and embarked Marines are permanently deployed in the Atlantic, Pacific, and Indian Oceans with the Sixth, Seventh, and Fifth Fleets, respectively. Disbanded in 1950, the Navy announced the re-establishment of the U.S. Fourth Fleet on July 1, 2008, which will operate in the U.S. Southern Command area of responsibility. The numbered fleets buttress security treaty obligations to NATO's current array of twenty-six countries and to Pacific allies—Japan, South Korea, Australia, and the Philippine Islands. In the Western Hemisphere, the Inter-American Treaty of Reciprocal Assistance (Rio Treaty or Rio Pact) includes most of the countries in the region, and it is rejuvenated annually by *Unitas* exercises undertaken by the U.S. Navy, often joined by U.S. Coast Guard and Marine Corps units, with Latin American navies.

The United States benefits significantly by forward deploying even so-called strategic weapons. Beginning in 1950, when sea-based aircraft became capable of carrying and delivering nuclear weapons, aircraft carriers had, in addition to their expeditionary role, a task as part of the U.S. strategic nuclear deterrent force. This was so important that the carriers that had acquired this capability—the USS *Midway*, USS *Coral Sea*, and USS *Franklin D. Roosevelt*—were stationed in the Atlantic and deployed to the Mediterranean during the Korean War. Put another way, the United States fought the Korean war, during which a substantial fraction of the airpower employed against North Korean and Chinese forces was sea-based, without the use of its three first-line aircraft carriers. Aircraft carriers were removed from the U.S. nuclear war plan—the Strategic Integrated Operating Plan—in 1976.[8]

Submarine-launched missile patrols began in late 1959. While forward-positioned, the function of these submarines was, and remains, deterrent rather than expeditionary. The first to deploy were diesel-electric-propelled submarines equipped with Regulus nuclear-tipped cruise missiles. The short range of the air-breathing Regulus missile required that patrols be conducted near the coastline of the Soviet Union, and in order to fire the missiles, which were carried externally, the submarine was required to operate on the ocean surface. In 1964 the Regulus submarine force in the Pacific was superseded by Polaris ballistic missile-carrying nuclear-powered submarines. Polaris missiles and their successors (the longer-range Poseidon and Trident missiles) could be launched while the submarine was submerged.

Ballistic missiles, deployed on submarines in broad ocean areas, greatly complicate an adversary's missile defense problem for two reasons. First, they have a shorter time of flight than ballistic missiles launched from the U.S. homeland, so the time available to react against them is less. Second, sea-launched missiles can approach targets from many azimuths, unlike land-based missiles that fly ballistic trajectories over the Arctic and approach through a narrow threat "tube." This can be seen on a globe by projecting trajectories from the "lower 48" United States to places on the Eurasian landmass.

Ballistic missile submarines have on occasion been employed in presence missions, as in 1963 when the USS *Sam Houston* on its first deployment to the Mediterranean made a port call in Turkey. The purpose was to allay Turkish doubts that U.S. submarines patrolling in the Mediterranean were a ready substitute for the Jupiter Intermediate Range Ballistic Missiles that were being removed from their deployment on Turkish soil.

Expeditionary sea-based ballistic missile defenses were proposed in the late 1960s. Positioned closer to an adversary's launch points, naval forces could effect early intercept of ballistic missiles in the ascending leg of their trajectories, when their speed is slower than during the reentry phase and their warheads and decoys would have not yet been deployed. This concept was called the "SABMIS" (Sea-Based Anti-Ballistic Missile System), but it remained only a concept and was not pursued.

Sea-based anti-ballistic missile capability is currently being embraced, however, for select U.S. cruisers and guided-missile destroyers equipped with AEGIS fire control systems. These ships will be used in expeditionary fashion to counter the threat posed by theater ballistic missiles by stationing them in international waters off the coasts of adversaries capable of launching ballistic missiles against U.S. allies.

U.S. strategists have known, like Mahan, that control of the seas was a vital mission with the purpose of keeping adversaries landlocked. To be able to use the seas to attain national objectives and prevent the adversary's doing the same was a precondition for success. The principle was rather simple, identical with that articulated in 1806 by Lord St. Vincent: "I do not say the French cannot come, I only say they cannot come by sea."[9] Geopolitically, U.S. presidential doctrines—from the Monroe Doctrine through the Truman Doctrine and the Carter and Reagan Doctrines to the current Bush Doctrine—have echoed the same principle: a strong desire to keep adversaries and potential adversaries as far from U.S. shores as possible. The forward operations of the U.S. Navy have always been, in large measure, for the purpose of lending operational muscle to these geostrategic policies.

Only from a forward posture could U.S. naval forces offer assurance to friends and allies, deter adversaries and potential adversaries, and be readily available to respond to crises, contingencies, natural disasters, or the wide variety of missions that might be required. Small forces in numerous places forward can gain important local knowledge, which then can become a valuable asset if a crisis should occur. Familiarity with the local environment facilitates intelligence gathering, reconnaissance, and surveillance, which are the key entering arguments for gaining access. "Access" is a recent term that encompasses finding an adversary and possessing the offensive capability to attack it, or to place forces ashore in a timely and effective manner.

Currently, expeditionary U.S. naval forces comprised of ships of all types and embarked Marines are permanently deployed globally. Cooperation and interworking with allied and friendly navies is routine. Chiefs of Naval Service exchange visits with the U.S. CNO (Chief of Naval Operations) on a regular basis. Indeed, from the biennial International Seapower Symposia first convened in 1969, to the establishment at the Naval War College of the Naval Command College in 1956 and the Naval Staff College in 1972, to the CNO's suggestion of a one-thousand-ship Navy in 2005 championing the active involvement of most of the world's navies, the U.S. Navy has been in the forefront of international cooperation for the freedom of the seas and the ability to use the seas in securing national interests. With a broad agenda of tasks to perform and long distances to cover forward deployed naval forces must be mobile, offensive, self-reliant, and adaptable.

Mobile

Garrisoned forces are static, for they are designed to hold or control territory. Because they are fixed, they tend to be predictable, for one can determine with some precision not only the origin, but also the limits and direction of their movements. Mobile naval forces, in contrast, are dynamic. Uncertainty as to their location, disposition, and movement is characteristic, and their freedom to sail unhindered around the globe underwrites that dynamism. Naval forces, including in particular the Marine Corps, do not want or like to sit still; for it compromises their sanctuary, which is enhanced greatly by mobility. Especially in view of the availability of overhead surveillance, high-speed communication links, high capacity computing, excellent locational abilities made possible by the satellite Global Positioning System, and accurate long-range weapons, to be stationary is to present a lucrative target. The impact and value of maritime mobility was succinctly stated by Bernard Oxman, when he wrote: "[T]he security of almost every state depends in some measure upon the mobility of the forces of naval powers for the maintenance of stability and security in its region"[10]

Historically, ships were used to transport troops and to mass them for invasion against undefended areas on coastlines. Before railroads and improved roads for cars, buses, trucks, and tanks, movement on land was much slower than movement of large, heavy units at sea. Advances in wartime of large military forces typically have been on the order of a hundred miles a week, if opposition was light. Speed of advance on land is hampered by the need for logistic forces to keep up—especially vital provisions of fuel and water. Seaborne forces can move on the order of four hundred miles a day, and they bring their logistic support with them. Putting this in perspective, one can imagine a U.S. amphibious task group in the Atlantic off New York City at noon undertaking a landing anywhere between northern Maine and southern North Carolina by lunchtime the next day.

High mobility coupled with the ability to elevate sensors into the atmosphere is vital to that most daunting of naval tasks: finding the adversary. Reconnaissance (finding) and surveillance (tracking) an adversarial force are the first steps to success in the open-ocean environment. Reconnaissance, surveillance, and reporting are the elements of what Wayne Hughes in his *Fleet Tactics* termed "scouting."[11] Recognizing these facts, naval forces strenuously avoid being detected themselves; if detected they seek to break track; and if tracked and identified, they energize layered defenses to protect themselves.

Since naval forces at sea are always in motion, whenever they are undertaking a mission they are *maneuvering*. Maneuver, properly understood, is not movement or mobility. Maneuver is relational: movement to place forces in a position of advantage with respect to an adversary—whether for offensive operations or defensively to avoid detection or render defensive systems more effective. Maneuver in itself makes no independent contribution to success. The great professional boxer Muhammad Ali used to boast: "Float like a butterfly, sting like a bee." It's not the float that matters, it's the sting. It is maneuver combined with attack or the threat of attack, or maneuver combined with defense that works to produce the desired effect. In land warfare maneuver is a variable—an option; in the maritime arena, maneuver is a constant—a way of life.

Offensive

Taking offensive action is necessary to victory in warfare. If one's defenses are impenetrable, then one probably cannot lose; but it requires offensive action to win. If one's objectives are merely not to lose, then offensive capability is shunned, as in the movement in the 1980s in Europe to embrace a "defensive defense" against the prospect of Soviet aggression. But, as French strategist Andre Beaufre cautioned: "The defensive can only pay if it leads sooner or later to a resumption of the initiative, in other words to some offensive action. A counter-offensive is essential if submission to the will of the opponent is to be avoided."[12]

The concern expressed by those who championed a defensive defense was that merely fielding offensive forces at best provokes the other side to acquire offenses of its own, and at worst, to attack preemptively. If, on the other hand, one cares about prevailing in wartime, offensive forces are required. Bernard Brodie put it this way: "Regardless of how impregnable our defenses are, and they can never be perfectly so, our security in the world is jeopardized if we cannot inflict vital injury on any nation which menaces us or our legitimate interests. And for this our Navy is indispensable."[13]

The offensive is all about seizing, retaining, and exploiting the initiative. This is effected by acquisition of powerful offensive weapons and by ensuring that one maintains freedom of action to deploy and to employ them. In this case the allusion is to using the Navy for offensive power projection operations—attacks by missiles and aircraft against landward targets from the sea, or the undertaking of amphibious operations. It is through the use of offensive capability, or the threat to employ it, that naval forces draw their leverage to shape political-military environments on a global scale. It has been pointed out, time and again, that the great majority of people in the world live within two hundred miles of the sea—and those numbers are increasing. The reason that this is an important statistic is that if military force is to be used from seaward axes, large numbers of people and organizations could be at risk, and this has political consequences.

For naval forces, the offensive is also about a stout defense being based upon an aggressive offense, or what has been termed "attacking at the source." Once a ship or submarine—or an aircraft, for that matter—leaves its landward sanctuary and moves to the great oceanic commons, it enters the realm in which the most difficult problem for opposing forces is to find it. Consequently, the attacker would prefer to strike the opposing naval force in its geographically fixed port or airfield. Admiral Forrest Sherman framed the issue as a matter of defending at-sea forces in this way: "The worst place to protect a ship is where the ship is. The worst place to protect a convoy is at the convoy. The worst place to protect a city from air attack is at the city. The best place is at the bases from which the airplane or the submarine comes. The next best is en route—the worst place is at the target."[14] Time and again this stands out as a recurrent theme in U.S. naval strategy and operations. As one study remarked: "[T]he U.S. planning objective of offense-in-depth . . . continued to dominate Navy thinking. . . . It outlined air strikes against submarines at their home bases; air and submarine mining of Soviet ports and training and transit areas. . . . U.S. naval air and missile strike capability was viewed by the Navy as the most effective way of reducing the enemy submarine force before it left port."[15] To undertake military attacks against the shore is what a forward-deployed, offensive-capable Navy is all about—both as a matter both of attaining political-military objectives, and as a matter of force protection.

It happens, moreover, that offensive actions against targets at sea, not only on the land, is the key to success. It has long been an axiom of land warfare that the defensive is the stronger form of war. Theorists of land combat have suggested that the defense is three or more times stronger than the offense, arguing that offensive forces, to be successful, must be three or more times larger than the defenses they attack. Yet, at sea, the reverse is true. In the maritime realm the offensive is the stronger form of warfare, for all those conditions that on land tend to strengthen the defense vis-à-vis the offense are absent—for example: large numbers of forces, fortifications, natural barriers, and havens.

Attacks against ships in the blue-water environment can originate from anywhere—the surface of the sea, beneath the sea, the air, the land, or even from space. The characteristics of attacks originating from different places can be markedly different, and they tend to place great stress on defensive systems to protect against them. Physical refuge, such as that provided by topographical features and man-made structures on land, are virtually non-existent at sea. Likewise, there are no fortified geographic strong points in the open ocean environment, and while the ocean depths do provide a significant measure of freedom from detection for submarines, the increased ambient pressure makes them more vulnerable to attack.

Great expanses of ocean offer the opportunity for ships to avoid contact with others, if they so desire. This, along with the absence of physical strong points at sea from which to mount a staunch defense and the fact that any state's inventory of warships tends to be small—which makes the value of each quite large—leads to a conclusion that it is crucial in open-ocean warfare to attack first and attack effectively. "Naval combat [in the open ocean] is a force-on-force process tending, in the threat or realization, toward the simultaneous attrition of both sides," wrote Wayne Hughes in *Fleet Tactics*. He continued: "To achieve victory one must attack effectively first."[16] All the reasons that make the defensive stronger on land are absent in the maritime environment. Sea battles through the ages confirm the wisdom of this factor in blue-water warfare.

An offensive mind-set, to attack adversaries "at the source" and to "attack effectively first" at sea is pivotal to the success of naval operations, and the U.S. Navy has embraced it from its inception. Clearly, attitudes that encompass this offensive approach lead to a tradition based on great victories in battles and campaigns at sea, fostering high espirit de corps and self-confidence in the force. It is important to note that this is not just the result of a Mahanian fascination with great sea battles as some critics suggest. Rather, it is demonstrably the way to win at sea, whether in a tactical engagement between two small units or a decisive battle between massive fleets.

Self-Reliant

In far forward zones of activity, ships might not have the benefit of support from other U.S. forces, from allies, or from friendly states nearby. U.S. Army and Air Forces must maintain bases from which to operate; naval forces act as their own seaborne bases. Because logistical support and overflight rights can be problematic, ships must be fully self-reliant. U.S. warships are designed to have extended unrefueled operational reach, so that they need not be dependent on fueling, repair, or victualing from ashore. Their mobility means either that the sources of resupply can come to them, or that they can transit to acquire what they need. Combatant ships of frigate size or larger routinely refuel and replenish at sea, essentially freeing them from particular shore bases. Since World War II the Navy has maintained large replenishment ships; no other Navy maintains and so highly values such a large, globally deployed capability.

Except for fuel and ammunition, fossil fuel–powered ships normally operate with at least two months' supply of provisions and spare parts for installed systems. As with sailing ships, the sustainability of nuclear-powered ships, on the other hand, is limited by concerns for the well-being of the crew. In World War II Germany conducted some limited reprovisioning of its submarine crews with so-called "milk cow" submarine tenders. At-sea transfers of cargo to modern nuclear-powered submarines is extremely difficult and far from routine, largely due to their lack of stability while on the surface.

Expeditionary forces, since the 1930s, are task organized. They are formed into self-reliant expeditionary task forces, task units, and task elements. Expeditionary tasks not involving naval forces can be expensive—both in economic and in political terms. Because naval forces do not need to ask permission of anyone to go anywhere—including in the territorial waters of other states so long as they are conducting innocent passage—because they are self-contained, and because they tend to be out-of-sight out-of-mind, the political and economic costs attending their expeditionary nature tend to be low.

Self-reliance, enhanced by the lack of political controls and the relative absence of geographic obstacles in the oceanic environment, is abetted by uncertainties involving an adversary's disposition and movement. At sea, naval commanders need not be constantly looking over their shoulder for support or "reaching back" for assistance. This permits them to look forward continually, always thinking and planning ahead.

Self-reliance is underwritten by multi-mission capability. The need to endow ships, even submarines, with self-sufficiency, multi-mission capability, and long legs requires that they be large in size. The aircraft carriers operated by the U.S.

Navy since World War II are among the largest mobile machines the world has ever known. Their size has been rather rigidly determined by the need to operate high-performance aircraft from them, and by economies of scale. Steel is cheap and, roughly speaking, for a 25 percent increase in the size of an aircraft carrier, the ship can house and operate 30–40 percent more aircraft and store the fuel to fly them. Together, all of these characteristics—the lack of political controls, the unencumbered oceanic environment, the independence from daily logistical support, the lengthy on-station sustainability, and multi-mission capability—foster a highly imaginative and innovative mind-set in the operators of the Navy.

Ballistic missile submarines communicate with other stations only rarely. They are self-reliant to the point where they can launch their ballistic missiles without any electromagnetic emissions from the ship beforehand. Command and control is exercised from afar; the "go" signal can be received by the submarine, even submerged, and acted upon without transmissions of any kind by the ship. Likewise, rules of engagement permit defensive use of military force by commanders of surface ships—even actions based only on the "hostile intent" of an adversary—without reference to higher authority. This was necessary owing to the Soviet doctrine of "the struggle for the first salvo," as a result of which Soviet forces were configured for a first strike against the U.S. Navy, primarily its aircraft carriers.[17]

When the Soviet Union was the major adversary of the United States, the navies of the two sides worked out an agreement, called the "Incidents at Sea Agreement," that sought to prevent incidents arising out of ambiguous or deliberately provocative acts. A code of conduct for operations on the high seas was struck in the late 1960s and early 1970s that not only survives to this day as a navy-to-navy agreement, but also offered a template for other navies to emulate. The bilateral Incidents at Sea Agreement (Technically, the U.S.-U.S.S.R. Agreement of the Prevention of Incidents On and Over the High Seas) prohibits, for example, the simulation of attacks on or launching objects at ships of the other signatory. Representatives of the involved navies meet on at least an annual basis to review the agreement. Not only does this represent an example of the Navy's desire to remain politically unencumbered in its oceanic operations, but also it offers an example of the kinds of actions that can be accomplished in a strictly military-to-military manner to lessen tensions and tamp down the potential sparks of conflict. Even more, the tenets of the agreement have been absorbed into the Navy Strategic Culture: today ship and organizational commanders would not even consider the kinds of actions that are proscribed by the agreement. It stands as a stellar example of an enduring change to the culture, much for the better.

Adaptable

Expeditionary U.S. naval forces are forward, mobile, offensive, and self-reliant. These are operational characteristics. One more operational characteristic is very important to successful operations in peacetime, crisis, and in armed combat, and that is adaptability.

Adaptability is the ability to adjust to the current environment or context of operations in order to achieve success. Adaptability is underwritten by flexibility, for flexibility is a property built into platforms and weapon systems, doctrine, plans, and the modes of thinking of decision makers at all levels. Flexibility is designed into systems—either technological systems or thought systems. Adaptability requires knowing about the flexibility at one's command, and how and when to take advantage of that flexibility.

Adaptability is a measure of how well one can cope with the exigencies of the environment, with the "feedback" from the system, given the flexibility offered by the systems at hand. A 5-inch gun mount on a destroyer has many built-in levels of flexibility of control and operation. It can be centrally and remotely controlled by the fire control system of the ship. It can be locally controlled within the mount. It can be operated remotely or locally. In older, World War II models, the 5-inch guns could even be "laid"—trained, elevated, and fired—in "hand" control; that is, with no electrical power supplied to the mount.

Flexibility increases the options available; adaptability is selecting or devising the right option at the right time. Adaptability is about context. A significant ability to adapt not only enables success, but at the same time it helps to tailor ways and means more closely to ends, which is another way of saying it decreases operational risk.

Flexibility is increased by quantity: the more weapons in the inventory, the more adaptively they can be used. Flexibility is also a function of multi-mission capability: if one has no amphibious forces, for example, that mode of warfare is not an option. Indeed, in the late 1960s the "Fast Deployment Logistics" force was proposed and then ultimately scotched as a concept in the Congress. Suggested as a force to respond "to any problem at any spot on the globe" with "discrimination and speed," it envisioned forces on the order of six squadrons of heavy lift C-5A, fourteen C-141 aircraft, thirteen FDL ships, prepositioned materiel and equipment in Europe and the Pacific, a Civil Reserve Air Fleet, and 460 general purpose cargo ships to be used out of the Military Sea Transport Service. The program was defeated and abandoned because there were those in the Congress who argued that such a force would constitute a temptation too difficult to turn away from and, as a consequence, would cast the United States in the role of global policeman. A perceived notion that presidents would be tempted to intervene "just because they could" swayed many in Congress away from the concept. George Washington had penned an apt response to this idea nearly two hundred years before, when he wrote: "No man is a warmer advocate for proper restraints and wholesome checks in every department of government than I am; but I have never yet been able to discover the propriety of placing it absolutely out of the power of men to render essential services, because a possibility remains of their doing ill."[18]

Flexibility is promoted by information systems that provide key information at the right time. The ability to exploit the effects of uncertainty and time is enabled by flexible command and control systems that are information-rich. Greater flexibility in command and control equates also to a decrease in centralized direction. It is a fact of life that expeditionary naval forces—those that are forward, offensive, mobile, and self-reliant—require greater flexibility, and concurrently the ability to adapt to conditions in the battlespace. As Arleigh Burke once put it: "We [the Navy] decentralize and capitalize on the capabilities of our individual people rather than centralize and make automatons of them. This builds that essential pride of service and sense of accomplishment."[19] Decentralizing requires that those on whom decisions are conferred have an open, but also an educated mind: "educated" in the sense of understanding the operational context in all its complexity and with historical insight, the actions of the adversary, and the range of options available.

The U.S. Navy has long been imbued with the ability to adapt. Adversaries can be seen to find this approach to military operations useful for them as well. A book published in 2006 offers an approach to a "management philosophy" that is familiar to naval officers: "Al-Qaeda had developed a management philosophy that it called 'centralization of decision and decentralization of execution.'"[20]

One might question whether flexibility, and its operational partner, adaptability, is really something one needs, or whether it might just increase temptations and risk-taking. Many reasons argue, persuasively, for investing systems and decision makers with flexibility so that they are able to perform adaptively. In the first place, pre-conflict assumptions might be erroneous. Assumptions must be made about future conflict: about the capabilities and also the intentions of adversaries; about the relevance and thoroughness of training; and about the impact and predictability of technological advances, for example. As often as not those assumptions are incorrect.

Secondly, as Colin Gray has so often warned, real wars are not the same as paper wars. One can write scenarios, conduct games and exercises, and prepare the best campaign plans on paper, but one must understand that these efforts do not necessarily conform with what happens when swords are actually crossed and the bullets and missiles fly. Military conflicts never follow pre-scripted courses. Good strategists always try to play a larger hand than they hold—to play on the uncertainties and fears of adversaries and convince them that their abilities will not be sufficient to prevail. Moreover, the real enemy is never the same as the paper enemy. One can study enemies for many years and never really know them. During the Cold War the United States had legions of government employees in multiple intelligence agencies and faculties of colleges and universities focused strongly and narrowly on the Soviet Union. None were confidently able to predict even system-shattering events like the fall of the Berlin Wall or the non-violent demise of the Soviet Union. The adversary is an independent agent capable of making autonomous decisions directly affecting conflicts in ways that are difficult to foresee or prepare for. That tends to be deliberate: operational and technical surprise are an important feature of warfare that, by definition, cannot adequately be anticipated. Nobody, but nobody, can write down a list of all the things that would never occur to him or her.

Third, weapons might not work the way they were designed, or doctrine for their use might be faulty, and need to be revised owing to the fire of combat. It happens that quite often weapons are designed and fielded, but the actual clash of arms is required to assess and evaluate their effectiveness. Exercises, simulations, and experiments can create the reality of the battlespace only up to a point. Many weapons fielded in armed forces around the world have no history of use in actual combat. They might perform up to expectations, they might perform above expectations, and they might fail completely like U.S. torpedoes early in World War II in the Pacific. Flexibility in the form of weapon redundancy, multi-mission capability, innovation in use, and large inventories helps to offset those kinds of problems.

Fourth, by offering more options flexibility helps to mitigate risks. There are no winning risk-free approaches to the use of military force; and, to boot, risks are very difficult to assess with any degree of accuracy. Furthermore, U.S. strategists are heavily hampered in ways and means by a myriad of organizational, operational, legal, and moral constraints.[21]

Adaptability facilitates the rapid transition from benign peacetime cruising to full-up hot combat operations. Because of their inherent flexibility fostering adaptation to new circumstances, naval forces are vital for coping with the dynamics of this difficult transition. Expeditionary naval forces transition most easily, because except for ordnance actually being used, the environment is the same for them. Naval forces do little differently, other than being appreciably more alert and aware, and pulling the trigger.

The resilience of naval forces, hence their survivability, are enhanced by flexibility in all its forms. The elasticity to adapt and bounce back from a setback, for whatever reason, is an important feature of adaptability. Yet, adaptability is far more than reaction. It involves the tempering of anticipation as well. Forces that cultivate future options and the ability to execute them stand a better chance to survive and to prevail than forces that do not. "This is the Admiral Nelson Fire Poker Principle. Speaking with some of his officers the night before Trafalgar, Nelson picked up a poker and said: It doesn't matter where I put this—unless Bonaparte says I must put it there. In that case, I must put it someplace else."[22]

Whether in modes of hardware, software, or thinking, adaptability requires knowledge of what flexibility is possible, what is permitted, and the imagination to go beyond what is believed possible and permitted. This has been called "thinking outside the box," when the box comprises what appears possible and permitted. Inflexibility of systems or in thinking creates boxes with walls that are strong and high. The open ocean environment, and the manner in which U.S. naval forces have been trained and operated for more than two centuries, temper the strategic culture and break down the boxes.

Expeditionary naval forces have long been forward, mobile, offensive, self-reliant, and adaptable. These are their strengths, and they have both appeal and applicability to the future. The next chapter will consider more carefully how, in addition to these factors, technology and systems thinking feed into and condition Navy Strategic Culture.

CHAPTER 6

Technology and Systems

Long has it been asserted that the Army equips the man, while the Navy mans the equipment. And long has it been true. That the Navy has had a lasting love affair with technology should come as no surprise to anyone. The reason is clear: technology represents life. It keeps surface sailors alive out of the water, where they want to be; and it keeps submariners alive under the water, where *they* want to be. Technology, however, is more to mariners than life: it is home. To be sure, the airmen of the Air Force and the tankers of the Army are at one with their technology, and it is vital to their survival. But they do not live in it. Sailors do.

Naval technology is complex because the environment is complex and difficult to master. Moreover, the risks and costs of failing either in weather or battle are ultimate. So, one finds, with good reason, that "an ocean-going ship, with her masts and sails, was incomparably the most elaborate mechanism which the mind of man had yet developed."[1] This constitutes a truism easily projected to the present to describe the most complicated very large moving object on the face of the earth: the modern-day aircraft carrier.

From the onset, Yankee ingenuity in seafaring—in its broadest sense—was both stimulated and guided by the British. Drawing upon a far larger and varied industrial base, and populated with artisans in seafaring trades, the Royal Navy stood as a formidable competitor, and model, to the upstart Americans. From the beginning—from *Old Ironsides* onward—U.S. naval forces sailed in high-quality, powerful, seaworthy ships. Moreover, their relationship to naval culture was close:

"The U.S. Navy between 1797 and 1860 actually created a tradition of progressive warship design while reinforcing desired continuities in American naval culture."[2] From this running start, the United States rapidly proved itself not only as a maritime rival to the British, but subsequently as a world-class innovator in armored ships, submarines, nuclear power for propulsion, aircraft carriers, sea-based aviation systems, ship-launched missiles, and a host of sub-systems, from avionics to navigation to oxygen generation to sonars.

In the era of sailing ships, speed and endurance melded with a heavy weight of broadside firepower to produce first-rate navies. Because wooden ships were exceptionally difficult to sink with solid shot, the ability to grapple and board and then to prevail in hand-to-hand combat was also highly prized. (The routine use of exploding projectiles came about only after the appearance of iron-hulled ships.) Speed helped to solve that chronic central problem of maritime warfare: finding the adversary. And speed was also key to avoiding battle if that was what one opted to do.

Then, as now, size was an important factor underwriting two of the most important characteristics of combatant ships: firepower and endurance, both of which were required if one wanted to operate in expeditionary fashion, or merely to patrol at long distances from friendly ports. Larger ships could produce more devastating broadsides, travel farther, and remain on guard at sea—on blockade station, for example—longer than their smaller counterparts. Because they could carry more armor, could absorb more punishment, were more highly compartmented, and—recently—could operate aircraft in heavier seas, they had higher expectations of survival. It is important to note in this regard that no *Essex*-class aircraft carrier—the class of large U.S. carrier that began service in 1942—was lost during World War II (or since, for that matter). Furthermore, no aircraft carrier since World War II has been lost to battle damage, accident, or weather. During the Vietnam War, in 1969, USS *Enterprise*, the only nuclear-powered aircraft carrier at the time, withstood the detonation of nine large bombs in an onboard fire. The ship not only survived, but could have resumed air operations within one day.

For monohull surface ships, speed and size are not closely related. ("Monohull" ships have a single hull. This distinguishes them from multiple hull ships, generally called catamarans or trimarans.) Large monohull ships can proceed nearly as fast across the surface of the water as small ones. The hydrodynamics of hulls and screw propellers pose a practical upper limit on surface ship speed because the power required to move a ship through the water is approximately proportional to the cube of the speed of the ship. What this means is that an upper speed limit exists for monohull, screw propeller–driven surface ships, because increments of increased power do not result in correspondingly higher speed capability. A doubling of ship's speed requires an application of about eight times more power. While the practical upper speed limit is different for submerged submarines because they do not have to deal with the friction of the sea-air interface, a similar power-speed principle applies.

Nuclear power enabled submarines to become true submersibles, capable of circumnavigating the earth without having to surface. It increased by orders of magnitude not only the unrefueled operational range of nuclear-powered ships, but also the striking power of large aircraft carriers by allowing virtually all of the more than three million gallons of fuel carried onboard to be used to power aircraft of the embarked air wing. Starting in 1955, with USS *Nautilus*'s signal: "Underway on nuclear power," the Navy rapidly became the world's leader in ship nuclear propulsion development, the impact of which was succinctly captured in this assertion: "Never has a naval engineering project of such complexity (the development of nuclear propulsion, with the concurrent evolution of the nuclear submarine) been accomplished successfully in so short a time."[3]

With the knowledge that ships would operate for long periods of time without replenishment, and without access to repair facilities, the installed equipment—from small relief valves to large turbines and reduction gears—had to be expertly designed, constructed, and operated. This offers stark contrast to what have been called "daylight navies" that cruise only in close proximity to their home ports and infrequently operate their diesel-electric submarines, and even less frequently submerge them.

Large ships take advantage of economies of scale. Steel for shipbuilding is inexpensive compared to the cost of crewmembers, so doubling a ship's size increases the required crew size by only about 30 percent, while increasing fuel requirements (for non-nuclear-powered ships) by only about 20 percent. As was pointed out in Chapter 5, large aircraft carriers also carry proportionately more aircraft than smaller ones.

In the era of sailing ships, as now, it was technology that bound the sailor to his unit. "Sailors," wrote Clark Reynolds, "learn their professional skills at sea among a small ship's company, where administrative and political considerations are minimal. They are technical experts, skilled in the technology of service at sea and sensitive to the inherent fragility of their machines."[4] On-the-job training rather than doctrine was employed to build individual competency.

Teams within ships were knit together by tasks. Where teamwork was required, written procedures acted as indispensable guides. Nowhere has this been more evident than in damage control, the crucial survival skill for all ships.

In general, ships fought as individual units. While ships of the line could be coordinated in their action by visual means with flag hoist and signal light, it was by far the exception rather than the rule that ships acted synergistically in battle. One special case was interactive participation in large amphibious operations. Ships often pool their defensive resources to protect one another, but in offensive operations they have acted collectively primarily in attacking targets ashore—strike and amphibious warfare. Major strike capability with aircraft and missiles has come to naval warfare only since World War II, with the advent of long-range attack aircraft and missiles, augmented more recently with computers and high speed information links.

Systems

In the maritime realm, *technology* is organized into *systems*. Systems dominate shipboard life: electrical systems, fuel systems, firefighting systems, propulsion systems, fire control systems (for the management and direction of the ship's weapons), and thousands of others. Accordingly, Navy Strategic Culture is suffused with systems thinking. Systems have structure and organization, they function in a given way, and the parts of systems have specific relationships one to the other—functional as well as physical. In linear, physical systems, inputs and outputs are related and can be analyzed and, for the most part, measured with accuracy.

Systems are reliable to the extent that they are well designed, well engineered, and have built-in redundancy. Because of the possibility of damage by high seas, foul weather, or combat, shipborne systems require that survivability be a key factor in their design. Whether the issue concerns routing of connecting cables between parts of a system redundantly through different areas of the ship, providing for modes of operation of the system under varying conditions of electrical power availability, or designing and operating the system to fail gracefully (that is, able to perform at successive lower levels of effectiveness after suffering accumulating damage) naval systems are designed with characteristics that ensure their survivability and continued effectiveness under conditions of extreme stress.

The number of systems that comprise naval warfare is very large, and their composition is varied. Systems are employed for directing the fire of guns, ensuring the availability of spare parts for equipment, managing personnel, sharing and evaluating information, feeding the ship's crew, delivering the mail, dewatering compartments after battle damage, or for hundreds—indeed thousands—of other purposes. They are everywhere and touch everything in ships. As a consequence, naval officers become quickly accustomed to thinking in system terms.

Combat systems may be classified as active or passive. Active systems project force or radiate energy; passive systems detect active system emanations. Active systems can perform "hard kill" of targets by depositing kinetic energy on them—hitting a bomber with a surface-to-air missile, for example—or "soft kill" as in the deflection of an incoming missile by means of electronic countermeasures. Active systems transmit energy and home on the return of reflections of that energy from their targets; passive homing systems detect and act on emanations by their target—noise, heat, light, physical contact, or electromagnetic energy. Passive systems require a "cooperative" target; that is, the target must either physically approach the passive system (such as sea mines), or it must undertake some positive action that can be detected by the passive sensor.

The relationship between active and passive systems, and their association in turn to the core problem of finding the adversary can be captured in a simple metaphor: two very strong men, enemies, are in a totally dark room the size of a football field. Each has the mission of attacking and defeating the other. They have one weapon: a baseball bat, all their natural passive sensors (sight, hearing, touch, and smell), and one active sensor: a flashlight. Use of the active sensor, the flashlight, to find the adversary will obviously reveal one's own position. How this contest will terminate will depend on the creative use by each of his sensors—especially his passive ones—or perhaps of deception and surprise. Technology's influence in this simple example would be to make available to one of the contestants better weapons, sensors, or information than the other.[5]

Techniques for attacking and defending against active and passive combat systems differ, but they must be considered in the context of systems. That is, one must consider the various ways to approach or to defend against active and passive systems by considering first how the system works, then how the individual parts operate within the system, in order to know how best to apply countermeasures. In Chapters 4 and 5, for example, the argument was made that the best way to attack adversary submarines is while they are in port, before they can disperse at sea. Thus, the system of submarine warfare is optimally attacked not at sea, but in the home ports of the submarines. If, for example, the submarines of an adversary are berthed in a harbor that has a bridge across its narrow exit channel, attacking the bridge so that it falls into and blocks the channel might constitute the best method to bottle up the submarines. Next, if submarines are already at sea, one would attempt to prevent their passing a particular choke point. In the Cold War, NATO had planned to patrol and fortify the line between Greenland, Iceland, and the United Kingdom (called the "G-I-UK Gap") in order to interdict the passage of Soviet submarines into the Atlantic Ocean. This is a hypothetical example of the "forward" part of an anti-submarine system, one that has many layers of opposition for the adversary submarine to negotiate before it reaches its target on the open ocean.

Linear systems can be deconstructed; that is, they can be disassembled into their constituent parts for study and analysis. Outputs of linear systems are the result of the interaction of their parts. Linear systems can be disaggregated and then reintegrated with resultant outcomes that can be anticipated. Feedback in linear systems stays within the physical system, and it can be confidently modeled and accommodated.

Fleet Adm. William F. Halsey, commenting on operations in World War II, once said: ""A fleet is like a hand of cards at poker or bridge. You don't see it as aces and kings and deuces. You see it as a hand, a unit. You see a fleet as a unit, not carriers, battleships and destroyers. You don't play individual cards, you play the hand."[6] And, of course, as a good strategist, Halsey was always interested in encouraging the adversary to believe he held more and higher cards than he actually had. Indeed, in historical perspective, the tradition of referring to combat efforts as "systems," applying force in a systematic fashion, has an extended and admirable pedigree. Field Marshal Arthur Wellington, who defeated Napoleon, called his form of warfare a "system," as did Horatio Nelson before him.[7]

Yet, while shipboard mechanical systems are linear, and fully understandable in physical terms, warfare systems tend to be non-linear, and cannot be fully understood purely in physical terms. Feedback mechanisms in systems that are not wholly mechanical—like combat—can be totally unpredictable. Because connections between inputs and outputs of non-linear systems are not direct, and the feedback mechanisms do not adhere strictly to physical principles, systemic outputs may appear to be unrelated to inputs. An apparently small event can cause major repercussions in a short period of time; likewise, a major occurrence might have no appreciable near-term effect, and long-term outcomes may be impossible to assess.

In warfare systems the indirect connections far outnumber the direct, physical connections. Nevertheless, if one is accustomed to thinking in terms of linear systems, non-linear systems are more easily comprehended.[8]

Command and Control

Commanding and controlling naval systems—ships and aircraft task organized into elements, units, groups, forces, and fleets—has long been a highly contentious issue both within the Navy, and from the point of view of those in higher authority. For centuries, communication to and between ships was limited to line-of-sight—flags during daytime and light signals at night. Once beyond that powerful concept, the horizon, ships could not be directly controlled from ashore. Ships could literally pass one another at close range completely undetected at night and in daytime during times of low visibility. No one ashore could know with precision where ships were at any given time.

Controlling external emissions of all types, called "EMCON," was and continues to be an important method to thwart those who would penetrate the sanctuary provided to ships by the open seas. The lack of "connectedness" imposed by vast oceanic expanses, the curvature of the earth, and by a reluctance on the part of ships to use active transmitters was often sought after and even carefully cultivated, especially among submariners. It contributed to the cultural self-reliance of naval officers. Of course, darkness and inclement weather were also viewed as the friends of maritime sanctuary.

It was not until the early twentieth century that wireless telegraphy made its appearance, enabling communications to, from, and among ships of the fleet across greater than line-of-sight distances. The first combat employment of wireless radio took place in the Russo-Japanese War by Japanese cruisers that reported the approach of the Russian Fleet, and alerted Japanese Admiral Togo, who then smashed the Russians at the Battle of Tsushima in May 1905.

The power of information to the system lies in its sharing, not in its hoarding. This explains why wireless communications were so important to navies: they could now "see" over the horizon because a forward scout could share its information, in real time, with fleet units and headquarters far away. At the same time, certainly, the scout was revealing its existence, and perhaps its position, to those who could intercept its transmissions.

Despite greater and greater connectivity, the "out-of-sight, out-of-mind" tradition—the propensity to discount distant authority when it differed from the commander's perception of the local situation—persisted in naval culture. It often led to conflict with superiors, and fueled their frustration. This was highlighted by the reported exclamation of President Franklin Roosevelt: "To change anything in the Na-a-vy is like punching a feather bed. You punch it with your right and you punch it with your left until you are finally exhausted, and then you find the damn bed just as it was before you started punching!"[9] To be sure, the currents of "unless I receive specific orders to the contrary I consider myself to be on my own"—that is, "unless otherwise directed" or "UNODIR"—run deep and swift in Navy tradition.

While independence of action was prized and guarded, the need to link ships together and to share information—to develop methods to learn collectively and collaboratively what was "over the horizon," so that adversaries could be located and one could negate their sanctuaries by "attacking effectively first"—was put into action by the U.S. Navy early in World War II. Extending the perimeter of reconnaissance by aircraft scouts and radar pickets, and then sharing the information they gathered sparked the development—in late 1942 in the Pacific—of what became the shipboard hub of activity focused on the external environment, the Combat Information Center. These developments, the rudimentary beginnings of networking, aided significantly in defending against Japanese kamikaze attacks on the fleet later in the war, for no major surface combatant was sunk by a kamikaze.

The Naval Tactical Data System (NTDS) developed by the Navy in the early 1960s provided electronic data links between ships, and between ships and aircraft. It was "the first shipboard tactical data system in the world to use stored-program, solid-state digital computers. Also, the first to use multiple computers in a distributed tactical data processing system . . . [and] The first shipboard system in the world to use automatic computer-to-computer data exchange between ships and aircraft."[10] Thus, the computer and associated radio links became the catalyst for integrated data systems, permitting ships and aircraft to conduct high-speed information exchange in the next small step toward networked systems.

Over time, as naval warfare experienced concurrent changes in technical systems—increased threats as well as greater capability inherent in longer range, higher speed weapon systems, and a corresponding greater likelihood of information overload and command and control paralysis—the need to rely on decentralized decision-making grew significantly. Early on in these cascading trends, in the 1970s in fact, the Navy instituted the Composite Warfare Commander (CWC) concept in order better to cope with them. The basic CWC idea is simple: formally to decentralize the delegation of authority for combat actions to lower supporting echelons. Empowering subordinates to act without specific top level direction facilitated rapid reaction in complex, fast-paced combat environments. It permitted subordinate echelons to adapt quickly to changes in the context as they perceived it, and actually to anticipate change and act peremptorily. It was the license to self-organize, or "self-synchronize," without external command. Coupled with the concept of "command by negation"—a close relative to "UNODIR"—in which the subordinate commander, guided by a broad understanding of his superior commander's intentions, acts within his own significant discretion unless his superior specifically negates, or vetoes, his actions—the CWC concept made major strides to meet the altered, dynamic combat environment.

In this way, the CWC retains overall responsibility for and *command* of the force, but *control* is exercised by subordinate commanders acting individually and essentially independently. Subordinate commanders are organized around specific mission responsibilities, such as anti-air warfare, anti-submarine warfare, mine warfare, information warfare, and strike warfare. Aircraft carrier battle groups and amphibious task organizations operate with individual, separate CWC arrangements.

The ability to access and transfer large amounts of information, not just "data," but data upon which some processing has been performed to transform it into information, has paralleled important changes in the warfare environment. Modern military operations have become more and more complex, one of the manifestations (and causes) of which is the burgeoning availability of information. Operating in a world characterized by complex systems requires knowledge, agility, flexibility, and license, which together generate the ability to adapt rapidly to changing conditions.

As discussed earlier, contemporary battlespaces are non-linear in effects: the same input can produce different outcomes because the contexts, and feedback mechanisms within them, differ. It is not possible to deconstruct such a system in order to analyze its component parts. This is because dynamic interaction among the system's components is one of its defining characteristics. A picture of the complexity of the battlespace is made clear by this passage: "War is particularly complex when the targets are hidden, not only by features of the terrain like mountains or caves, but also by the difficulty of distinguishing among friends, enemies, and bystanders. It is also complex when the enemy is divided into diverse, versatile, and independent targets; the actions that need to be taken are specific, and the difference between right and wrong actions is subtle. Complex warfare is characterized by multiple small-scale hidden enemy forces. Large-scale warfare methods fail in a complex conflict."[11] From this it can be suggested that the non-linear, complex environment that naval officers have faced throughout history at sea, described in detail in Chapter 3, has now become commonplace on land as well.

Systems of Systems and Network-centric Operations

Just as NTDS began the structuring of systems into networks; just as the CWC concept and organizations adopted decentralization, delegated authority, and command by negation in order better to cope with the speed and complexity of modern combat operations; and just as the Navy's Cooperative Engagement Capability enabled suitably equipped ships to share both information and weapon systems at an extended distance from one another, the next steps were natural ones. It should come as no surprise that the concepts called "systems of systems" and "networkcentric warfare" should have been closely linked to Navy officers.

In his book, *Lifting the Fog of War*, Adm. Bill Owens wrote about "systems of systems," describing the systems as those associated with "seeing" (intelligence collection, surveillance, and reconnaissance), "telling" (command, control, communications, computers, and intelligence), and "acting" (application of precision force). Likewise, Vice Adm. Arthur Cebrowski led what was, in effect, an operationalization of the system of systems: network-centric warfare, or more generally, network-centric operations (NCO).[12]

Briefly, NCO generates increased combat power by networking intelligence sources, sensors, decision makers, and shooters to achieve shared awareness, increased speed of command, higher tempo of operations, greater lethality through the distribution of firepower, increased survivability, and a degree of self-synchronization. Information, in this scheme, moves rapidly through and across the chain of command. Thus, NCO can provide a clear, high resolution picture of the battlespace very rapidly. One obvious benefit of this is that it aids significantly to increase the speed of decision making. These are straightforward descendants of the CWC concept, which sought from its inception to share information in order to facilitate decision making and action at ever lower levels of the command structure.

One of the ways this is accomplished is by the distribution of forces into smaller force packages. Distributing military force can be used to lower the level in the command structure in which decisions are taken, while concurrently increasing both the surveillance requirements of the adversary and options for the friendly commander. Distributed capabilities can provide greater effectiveness and, consequently, fewer casualties.

NCO also takes combat operations two steps beyond the imposition of damage by exploding ordnance. So-called kinetic operations—steel-on-steel—have been joined in the information age by syntactic operations and semantic operations. The former, *syntactic* attacks, target the operating logic of systems. If, for example, a message can be diverted, or if it can be interrupted or altered in some way—not necessarily physically—then a syntactic attack can be said to have taken place. Syntactic attacks address neither the meaning of the message nor the purpose of the system: they work against a system's *functions* or *logic*.

In contrast, *semantic* operations target the meaning of information. Deception, psychological operations, and counter-operations security (counter-OPSEC) undertakings fall into this latter category. The intention is to deceive adversaries, or to undermine their trust in what their systems tell them. A syntactic attack would weaken confidence in the ability of a system to carry a message reliably to its destination; a semantic attack, in contrast, would seek to alter the meaning of the message, but not necessarily its transmission or reception.

The future battlespace will most likely witness greater use of syntactic and semantic efforts by all parties to the conflict. As a consequence, the increased use of deception, the premium on surprise, the rapidity with which targets can emerge and disappear, and the propensity for targets to be moving rather than fixed collectively point to the value of self-synchronization. This means that forces empowered with high situational awareness will be able to recognize, anticipate, adapt, and act without delay and without further direction.

Studies have demonstrated that at high levels of complexity and expertise, people do not even realize that they are making decisions. "Rather, they are fluidly interacting with the changing situation, responding to patterns they recognize."[13] Given this appreciation, three prerequisites must be in place for self-synchronization to take place and to succeed: first, there must be a body of doctrine to support the actions, and that doctrine (or as some have termed it, a "rule set") must be well understood and forces trained in its use; second, there must be communications among the units that self-synchronize to accomplish a common objective; and third, there must be a discrete commander's intent that provides the carefully considered conditions and rules under which self-synchronization may and may not occur.[14]

Networking is the means by which these important functions and capabilities are implemented. The result of all this is to translate combat power down to the lowest echelons, meaning that smaller forces can operate more quickly and effectively. This flattening of the command structure, abetted by increased automation of some operations, coincidentally means that higher authority can exercise greater and greater control at lower and lower levels. While it is true that this can increase meddling and inappropriate action from those who do not have the same picture as the on-scene commander, it is an unavoidable by-product of the need to share a common operational picture. As Michael Palmer wrote: "Thus the availability of advanced communications systems undermined but never completely destroyed the navy's tradition of reliance on 'the initiative of the subordinates.' The adoption of such systems allowed the higher echelons of command to severely limit the scope of initiative and, in many instances, to micromanage ongoing operations. With each passing decade, as the ability to communicate improved, the tendency to interfere grew stronger."[15] To some extent experience and wargaming of foreseeable situations can help to calibrate the intuitions of all, but the problem has no obvious solution.

Adversaries today also take advantage of networks and the technology that supports them. There is no reason to doubt that this trend will extend into the future. For current and prospective enemies, networks provide rapid, global means to recruit new members to their cause, to organize them, to transfer funds, to purchase the materiel of combat operations, or to spread propaganda—in brief, to effectively and efficiently share information in pursuit of their economic, social, military, and political goals. Indeed, "Insurgents have their own reconnaissance and surveillance networks. Because they usually blend well with the populace, insurgents can execute reconnaissance without easily being identified. They also have an early warning system composed of citizens who inform them of counterinsurgent movements. Identifying the techniques and weaknesses of enemy reconnaissance and surveillance enables commanders to detect signs of insurgent preparations and to surprise insurgents by neutralizing their early warning systems."[16] Clearly, such networks must be combated by networked counter-actions.

Networks countering networks are beset at the onset with problems of perception. From the inside, one's own networks often look weak and vulnerable. The same network viewed from the outside—through an adversary's eyes—however, might appear formidable, very elastic and nimble, and extremely difficult to degrade or disable. Well-designed and -operated networks tend to be models of flexibility and adaptability. In theory, networks are quite robust, and their design alone helps to thwart efforts to defeat them. A plethora of alternate routings, self-repair, and rapid reconstitution tend to be characteristic of resilient networks. Apart from the structure of the network, operation of the network by knowledgeable professionals provides the other ingredient of sturdiness. The key is to ensure that one's networks are both carefully designed, imaginatively secured, and skillfully operated.

Of interest, NCO is applicable at all levels of warfare—strategic, operational, and tactical, for it is transparent as to mission, force size and composition, and geography. Since NCO is oblivious to context, it acts as a means, or a comprehensive tool, that leverages technology and systems to support efforts in pursuit of its expeditionary, forward, mobile, offensive, self-reliant, adaptable goals.

Navy Strategic Culture, from the beginning cheek-by-jowl with technology, has undergone evolutionary modifications and transitioned easily into the twenty-first century's combat environment. Experiments and exercises, and actual combat operations will, over time, generate the learning and experience of how best to utilize these new, comprehensive, flexible means in order to attain objectives.

CHAPTER 7

The Maritime Strategy of the 1980s

In the 1980s the Navy originated and promulgated its "Maritime Strategy." The strategy, the account of which has been comprehensively detailed in a series of *Newport Papers* published by the Naval War College Press,[1] provided a rationale for naval forces and an overall strategy on how to conduct maritime campaigns in the event of a war with the Soviet Union. Not only did the Maritime Strategy act as a lodestar for planners, programmers, operators, and allies, but it represented a unique focusing of the various elements of Navy Strategic Culture.

As the naval component of U.S. National Military Strategy, the Maritime Strategy—revealed publicly in January 1986 in a publication of the U.S. Naval Institute[2]—embraced wartime expeditionary operations on the flanks of the Soviet Union, offensive assaults on both the Soviet fleet and the Russian landmass, and closely coordinated offensive and defensive actions with sister services and allies.

The *Maritime Strategy* began with an acknowledgment that there existed a "violent peace," one that demanded an operating tempo by the Navy in forward areas of the world's oceans 20 percent higher than it had been during the Vietnam War. Recognizing the potential threat of transnational terrorism—"by placing at risk forward-deployed forces, terrorists hope to be able to intimidate us into withdrawing, thereby undermining our credibility"—it presaged the attack on the USS *Cole* in the year 2000.

The Maritime Strategy envisioned forward expeditionary operations by the Navy in peacetime, complemented by a seamless transition in response to crises, and ultimately to a coalition war against the Soviet Union. Its authors presented the Maritime Strategy as an explanation of how the Navy and its alliance partners could "make a strategic difference" in a war against the Soviet Union. Previously, it was believed that the Navy's role would be peripheral and secondary, while the Army and Air Force together with the NATO allies would pursue the war effort in the center of Europe.

In asking themselves how the Navy could make a strategic difference in a war against the world's largest and most powerful land power, Navy strategists made a careful strategic assessment, tempered with an understanding of history and of geopolitics. The strategic assessment began with an understanding of the dimensions of warfare: *time*, *space*, and *intensity*. These are the variables that one seeks to control in order to prevail over an adversary. In the severe case in which the Soviet Union attacked NATO in the center of Europe with massive tank armies, the coalition strategy was to take advantage of all three dimensions in order to thwart Soviet intentions.

In *time*, NATO's strategy of "flexible response" sought to mount a strong enough defense against ground and air attack so that time would permit massive reinforcement and resupply by ship rapidly across the Atlantic Ocean from the United States—augmented by POMCUS (Prepositioning of Materiel Configured to Unit Sets) supplies stationed in-theater. Military shipping requirements amounted to on the order of 8 million tons of equipage and supplies and 15 million tons of petroleum products to be provided in the first six months of war. Economic cargoes would require an additional fifteen hundred ships to move on the order of 100 million tons of material per month.[3] Shipping availability and movement plans to and from ports, of course, had to be specified well in advance. This was unquestionably a vast, complex undertaking.

In *space*—meaning geographic terrain, not outer space—there was precious little flexibility. Germany was not neutral territory amenable to trading in order to gain time for reinforcements to arrive; the defense had to be mounted as far in the direction of the threat as possible: at the inter-German border. That defense must be stout enough to hold until reinforcements arrived or a decision could be made to escalate. And yet, politics intervened to prevent strong anti-tank defenses or barriers to be erected, for the West German government did not desire to place physical impediments in the way of eventual reunification with East Germany. The possibility of a powerful NATO counterattack on the ground in an eastward direction through a different geographic area to offset an offensive thrust by Warsaw Pact armies was also ruled out, for that would require placing offensive NATO forces in forward positions, which might look threatening to the Kremlin, and thus perhaps even precipitate a war. Likewise, redeployment of forces from their post–World War II locations to more defensible positions was nixed on political grounds.

In *intensity*, NATO's Flexible Response strategy included first a resolute conventional defense at the point of attack, even though NATO never achieved its stated force goals in forward ground and air forces. The option to use tactical (battlefield) nuclear weapons to frustrate aggression was not ruled out, and forces on the ground—and forward-based attack aircraft—were provided such weapons for their arsenals. From the onset and throughout the Cold War, this was highly controversial, for the European allies were deeply divided on the nuclear question. For example, considering the prospect of nuclear escalation that taken to its logical conclusion might include intercontinental-range nuclear weapons, France declared that the United States would never be willing to trade New York for Paris, and on this basis acquired its own nuclear deterrent, the "*force-de-frappe*"; and subsequently, in 1966, opted out of the military component of the NATO alliance. Some Germans resisted nuclear deployment in Germany of battlefield nuclear weapons, adopting "the shorter the range [nuclear weapon] the deader the German," as their mantra. Others in NATO countries insisted on a "defensive defense" only, with no prospects of offensive action either on the ground or in the air.

Having analyzed this situation, the leadership in the U.S. Office of the Secretary of Defense decided that the role for naval forces was to operate in the time dimension primarily, employing surface ships, submarines, and anti-submarine-capable aircraft to defend the reinforcement and resupply shipping. Navy strategists held a very different view, however.

Given the large numbers of air forces on both sides, the role that a comparatively small number of carrier aircraft could play early in the air battle over the continent was deemed to be insignificant. Moreover, foreseeing that amphibious landings on the flanks could be useful but not war-shaping in their potential outcomes, Navy strategists considered taking the war where the Soviet Union would not want to fight it—on the flanks of the Russian landmass. They understood that if the Soviet Union attacked NATO in Europe, that would be where the Kremlin preferred to fight. An essential element of the spatial dimension of strategy is to take the conflict somewhere that the enemy cannot ignore but does not want to engage. The flanks of the vast expanse of the Soviet Union offered that prospect. They reasoned further that strikes against the Soviet Union on its periphery, both in the Pacific and Atlantic (and Arctic) theaters would bring the war to the Soviet homeland, as the war in the center would not, and thereby stimulate a Russian defense, requiring them to deflect their strategic focus somewhere other than in the center of Europe.

Based on a long-standing predilection for "attacking at the source," Navy strategists reasoned that the Navy could defend the Atlantic sea lanes far forward by attacking Soviet submarines in their bases, and pinning them down in the Barents Sea so that they could not sortie into the Atlantic Ocean and interdict shipping there. With the assistance of Air Force fighter aircraft based in Iceland and an Air Force and Marine Corps Air presence inserted into northern Norway, moreover, they could establish an effective filter on the Soviet long-range aviation and Soviet naval aviation threats to open-ocean alliance shipping. They maintained that attacks against Soviet ballistic missile submarines patrolling in protected bastions in the Barents Sea and in the Pacific had the potential not only for pinning down Soviet defensive forces (surface ships, aircraft, and submarines charged with the mission of defending the ballistic missile submarine strategic reserve forces) in those remote areas (as discussed in Chapter 4), but also ultimately to attack the ballistic missile submarines themselves and cause a shift in the strategic correlation of forces unfavorable to the Soviet side. These were clearly strategic objectives, and campaign analyses conducted at the time indicated that they were obtainable, even if the risks were substantial.

Figure 2, from the presentation materials on the Maritime Strategy, depicts how the strategy sought by the Navy contributed to "make a strategic difference." This projection of the world was used consistently throughout the presentation of the strategy. The projection bisected the continental United States while keeping the Soviet Union integral, centered, and obviously landlocked. It emphasized not the sea areas, but the land areas—the seat of strategic effect. It portrayed the global nature of the conflict, not one confined to the central front in Europe. While both Atlantic and Pacific Oceans are shown in their entirety, they lie at the edges of the projection—at the periphery—demonstrating their value as highways to the scene of action, the Eurasian landmass. The Soviet Navy is squeezed in the calipers in its widely dispersed four fleet areas: the Northern Fleet on the southwestern Arctic Ocean (Barents Sea), the Baltic Fleet, the Black Sea Fleet, and the Pacific Fleet located in Vladivostok and Petropavlovsk on the Kamchatka Peninsula.

The Maritime Strategy was set forth in three phases: "Deterrence or the Transition to War," "Seizing the Initiative," and "Carrying the Fight to the Enemy." Attacking Soviet forces and the Soviet Union itself on its flanks was wholly in keeping with Navy Strategic Culture. The Maritime Strategy was expeditionary, forward, offensive, complementary with the assistance of other services and allies, and it permitted the adaptations that would be necessary to prevail. Recognizing

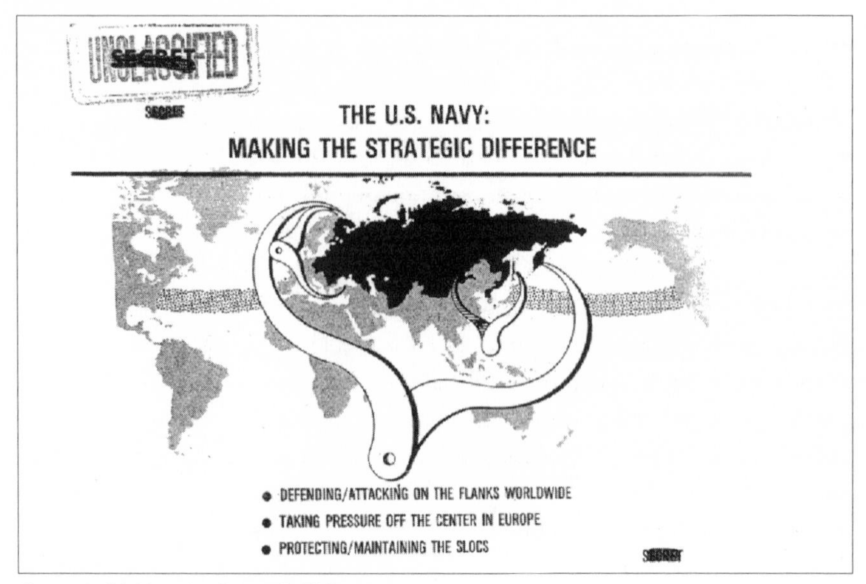

Figure 2. Making the Strategic Difference.

that ships are very survivable, the strategy sought to counter the opposing Soviet naval strategy by neutralizing it: destroying deployed ships, including submarines in their bastions, and pinning down the remainder in widely geographically separated home ports confined by very limited access to the open oceans. It sought to take advantage of a long naval tradition of excellence and victory at sea, while reinforcing any feelings of inferiority that the Soviet Navy might have, since it, unlike the U.S. Navy, had no history of the successful application of sea power in time of war.

Of interest, however, is not so much that the Maritime Strategy of the 1980s was in harmony with Navy Strategic Culture, for one should expect such for it to be embraced, placed into effect, and sustained by the officer corps and the civilian leadership of the Navy. Rather, it is important to understand how both national and naval historical interests and efforts were refocused, resulting in the formulation of the strategy in the early 1980s.

Recovering from World War II

In the aftermath of World War II, Army Air Corps Gen. H. J. Kerr wrote: "To maintain a five-ocean navy to fight a no-ocean opponent is a foolish waste of time, men, and resources."[4] This Air Force dogma centered explicitly on inter-service rivalry. "Why should we have a Navy at all?" said Gen. Carl A. Spaatz, "There

Table 2. Service Shares of the Defense Budget Compared

Height of Target	**1947–1971** (25 years)	**1972–1986** (15 years)
Navy Highest	Zero	13
Navy Lowest	10	Zero
Air Force Highest	19	2
Air Force Lowest	Zero	Zero

Derived from Department of Defense, *National Defense Budget Estimates for FY 2009.* Available at http://www.defenselink.mil/comptroller/defbudget/fy2009/index.html (October 22, 2008).

are no enemies for it to fight except apparently the Army Air Force."[5] These were among the opening salvoes in what was to bloom into bitter inter-service rivalry, inflamed by the standing up of the United States Air Force as a separate service in 1947, and the cancelling by the secretary of Defense of a new aircraft carrier, to be named the *United States*, in favor of a new Air Force bomber, the B-36. In fact, the secretary of Defense Louis Johnson was reported to have said in 1949: "The Navy is on the way out. There's no reason for having a Navy and Marine Corps. General Bradley tells me that amphibious operations are a thing of the past. We'll never have any more amphibious operations. That does away with the Marine Corps. And the Air Force can do anything the Navy can do nowadays, so that does away with the Navy."[6]

The "Revolt of the Admirals," ably and thoroughly chronicled by Jeffrey Barlow,[7] recounts this extraordinary period, marked by the simultaneous retrenchment of the U.S. military; the rapid rise of the U.S. Air Force accompanied by deep cuts in the Navy's budget; attempts to fashion a national strategy to deal with Soviet expansionist designs underwritten by the Soviet government's surprise explosion of an atomic weapon in 1949; and the formation in China of a communist government under Mao Zedong.

During the twenty-five years from 1947 to 1971, the Air Force claimed the highest share of the defense budget among the three services no fewer than nineteen times. In ten of those years, the Navy was saddled with the lowest share, while

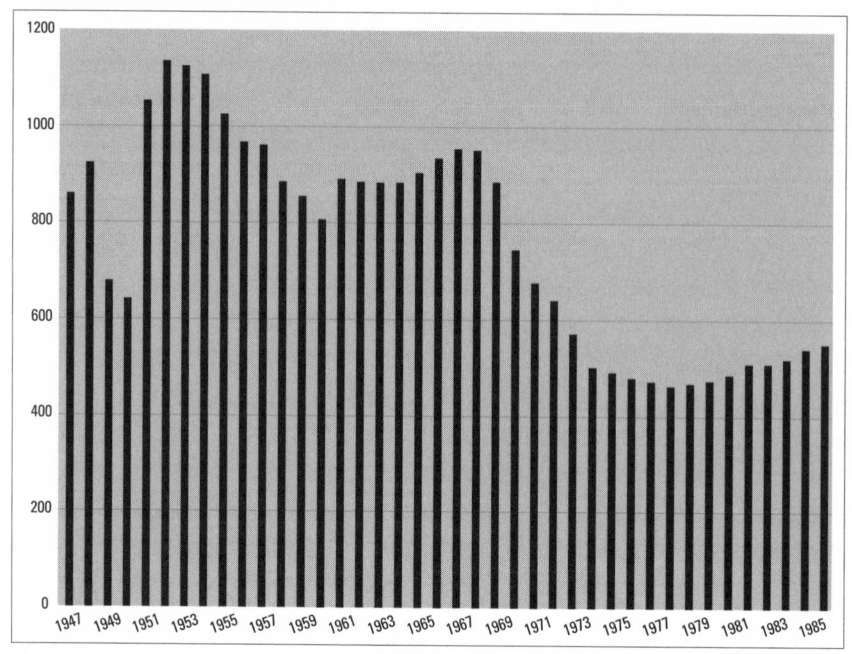

Figure 3. U.S. Navy Ship Inventory 1947–86.

the Air Force never found itself in last place (see Table 2). The general perception that each of the services receives about a third of the defense budget is not underwritten in fact, and when a particular service receives the largest share year after year, it is bound to cause friction and resentment among the others.

After 1972, however, the situation changed, and the Navy forged to the front with a long run of capturing the highest share of the defense budget among the services. This turnaround was the result of the confluence of several trends, the major ones of which were the need to recapitalize the fleet (a legacy primarily of ships built during World War II), the end of the building phase of over a thousand intercontinental ballistic missiles by the Air Force, and the growth of the Soviet Navy and large increase in its global operations.

Across this quarter-century, marked at its beginning by massive and rapid post–World War II disarmament and the roller-coaster funding of the Korean War, the Navy had done well owing to a large post–World War II inventory of combatant ships. By the 1970s, however, those ships were wearing out, and the drawdown was nearly complete. By the time U.S. fighting forces were completely withdrawn from Vietnam, 1973, the size of the fleet, at 576 ships, was approaching post-war low levels. It approximated only half of what it had been at the height of the Korean War. Figure 3 illustrates the changes graphically.

The Soviet Union as the Threat

Very soon after the end of World War II, the Navy identified the Soviet Union as the major future threat, the words of Secretary Johnson and Generals Knerr and Spaatz notwithstanding. In fact, the geopolitical specter of Russia dominating the Eurasian landmass had been written about by none other than Alfred Thayer Mahan nearly half a century earlier.

Mahan's words—not from his famous *The Influence of Sea Power on History 1660–1783*—but from *The Problem of Asia* are quoted at length because they were so influential in their future impact:

> Russia . . . is working, geographically, to the southward in Asia by both flanks, her centre covered by the mountains of Afghanistan and the deserts of eastern Turkestan and Mongolia. Nor is it possible, even if it were desired, to interfere with . . . this extended line . . . for the Russian centre cannot be broken. It is upon, and from, the flanks of this great line that restraint, if needed, must come.
>
> The struggle as arrayed will be between land power and sea power. The recognition that these two are the primary contestants does not ignore the circumstance that . . . the land power will try to reach the sea and to utilize it for its own ends, while the sea power must obtain support on land. . . . Hence ensues solidarity of interest between Germany, Great Britain, Japan, and the United States, [France was noted as the "conspicuous artificial exception" owing to her alliance with Russia] which bids fair to be more than momentary, because the conditions seem to be relatively permanent.
>
> Upon one flank of the Russian line lies the army of Japan; upon the other, five thousand miles away, that of Germany. . . . The two extremes of the Russian line, thus open to attack, are most inadequately connected by rail.
>
> From the conditions, we must be an effective naval force in the Pacific. We must similarly be an effective force on the Atlantic; not for the defence of our coasts primarily, or immediately, as is commonly thought—for in warfare, however much in defence of right, the navy is not immediately an instrument of defense but of offense.[8]

From these excerpts one can recognize a foreshadowing of the maritime strategy as it was articulated much later. Mahan establishes that Russia, *even before the Bolshevik revolution*, will be the primary adversary, that it must be opposed by *coalitions* of forces in both the Atlantic and Pacific, that coalition forces must take *forward positions* in ocean areas contiguous to Russia, and that strategic operations to oppose Russia from the sea must be *offensive* in nature and mounted on Russia's flanks, not in the center.

While in 1947 the Air Force generals were correct that there were no fleets to fight, it did not mean that the Navy was not needed. In January 1947, then Deputy-CNO Vice Adm. Forrest Sherman briefed President Truman on the probable character of a war in Europe initiated by the Soviet Union as follows:

> We envisage that from the naval point of view such a war would have four distinct phases. The first phase would be one of initial operations by our existing forces, of stabilization of the Soviet offensive, and of mobilization and preparation of additional forces, and of expansion of production of war material. The nation would be on the strategic defensive but our naval and air forces should assume the offensive immediately in order to secure our own sea communications, support our forces overseas, disrupt enemy operations, and force dissipation of enemy strength. In this phase the navy would have a tremendous initial responsibility. Early offensive blows would be of extreme importance in shortening the war.
>
> Strong submarine forces would be required for such tasks as destruction of enemy controlled shipping, reconnaissance, and inshore work, sea-air rescue, patrol of advanced areas and bottling up the Russian Navy.
>
> The second phase would be one of progressive reduction of Soviet war potential and build-up of our own. Operations would be characterized first by increased offensive action by naval and air forces and by joint forces, and subsequently by general advancement of our base areas as our military power permits. During this phase, large elements of all services would be moved overseas; advanced bases would be established and stocked; and requirements for shipping of all sorts and for naval escorts would increase rapidly.
>
> The third phase would involve a continued and sustained bombing offensive. Naval activity would consist of maintaining our overseas lines of communications, protection of troop movements, gunfire support for amphibious landings, carrier action against appropriate objectives, and submarine operations to prevent enemy use of coastal waters.
>
> The final phase would comprise the systematic destruction of Soviet industry, internal transportation systems, and general war potential. As naval targets disappeared, our naval operations would become more thoroughly integrated with ground and air operations, the need for maintenance of heavy carrier striking forces would decrease; while the need for ships for transporting forces and supplies, and for close-in escort and support would remain high.[9]

Virtually all of the threads of the Maritime Strategy of the 1980s are present in Admiral Sherman's presentation. When Admiral Sherman delivered it to the president, moreover, the Navy had the full wherewithal to perform precisely as he had outlined in his briefing.

In the Doldrums

Over the years, despite the bump-up in forces during the Korean War—especially in amphibious forces for the Inchon operation—the fleet declined in size, and budgetary support was not forthcoming to underwrite the aggressive offensive strategy outlined by Admiral Sherman. During the period 1955–1961, CNO Adm. Arleigh Burke strove vigorously to awaken the U.S. security establishment to the burgeoning power of the Soviet Navy and the threat it posed for the conduct of U.S. maritime operations. By the end of Burke's tenure in 1961 the attack carrier force was large and healthy, fleet ballistic missile submarines were being produced at an unprecedented rate, a large dedicated anti-submarine warfare force was in place to defend shipping destined for U.S. allies abroad, and amphibious forces were the cutting edge of U.S. policy in remote areas of the world. Yet, he had largely been unsuccessful in convincing the administrations under which he served to support naval programs in the manner he believed they warranted. The Navy played second budgetary fiddle to the Air Force throughout his tenure as CNO.

Although the U.S. blue-water fleet was obsolescent, it proved capable and proficient enough to support the land war in Vietnam. Strategy for use of the seas was virtually non-existent in that war because, even though an attack at sea (the Tonkin Gulf incident of August 1964) triggered greatly increased U.S. involvement, there was essentially no opposition to U.S. naval forces at sea.

The 1960s also witnessed a new emphasis on rationalizing and analyzing defense programs rather than on formulating and articulating rationales for the strategic use of military forces. Consumed by a war that struck hard at both morale and force structure, and compelled to meet analysts of the Office of the Secretary of Defense (OSD) in pitched battle on the budgetary front, maritime strategists in the United States retreated into almost total silence for better than a decade.

Focused so strongly on budgeteering, nuclear force posture, and war avoidance, OSD foresaw no offensive role for the Navy in a major war with the Soviet Union. Instead, it sought to emphasize the Navy's securing of the "Atlantic bridge" that would bring reinforcements and resupply to the war in the center of Europe. The Vice CNO, Horatio Rivero, characterized the attitude of the Office of the Secretary of Defense as: "Let's not have too many [carriers] and let's have as few other ships to go with them as possible. The Navy's real role is to convoy across the ocean [and ASW]." The Navy was not comfortable with this, characterizing it as "hauling ash and trash," which was, in a word, countercultural.

Adm. Elmo R. Zumwalt Jr. became the youngest Chief of Naval Operations in the summer of 1970. Zumwalt had been schooled in systems analysis, and understood the techniques that had been in vogue in the Office of the Secretary of Defense under the stewardship of Robert S. McNamara. Acknowledging that the strategic nuclear mission was in good shape across the board, and anticipating that increased funding for power projection forces would not be forthcoming from the administration, Zumwalt opted to emphasize the sea control mission. Power projection was rarely discussed as pertaining to conflict with the Soviet Union; it was to be used in *other* areas of the world in support of the Nixon Doctrine.

Invigoration

Nevertheless, in 1972 budgetary priorities favored the Navy, and for the next twelve years the Navy enjoyed the largest fraction of the defense budget among the services. By 1979, the new CNO—Adm. Thomas Hayward—could set forth his "Fundamental Principles of Naval Strategy":

> I am pleased to report that there is agreement within the U.S. government today on the proposition that maritime superiority must be the foundation of our national naval policy. . . .
>
> In conjunction with allies' maritime forces and facilities, our capabilities must be sufficient to put at risk the survivability of Soviet maritime forces even in their coastal waters and bases. . . .
>
> Fundamental to current naval strategy is the principle that U.S. Navy forces must be offensively capable. . . .We must fight on the terms which are most advantageous to us. This would require taking the war to the enemy's naval forces with the objective of achieving the earliest possible destruction of his capability to interfere with our use of the sea areas essential for support of our overseas forces and allies. . . .
>
> Keeping the Soviets preoccupied with defensive concerns locks up Soviet naval forces in areas close to the USSR, limiting their availability for campaigns against the SLOCs, or for operations in support of offensive thrusts on the flanks of NATO, or elsewhere such as in the Middle East or in Asia.[10]

When the Reagan Administration was installed in early 1981, its new secretary of the Navy, John F. Lehman Jr., sought to implement the Republican Party platform for the 1980 elections that stipulated: "Republicans pledge to reverse Mr. Carter's dismantling of U.S. naval and Marine forces. We will restore our fleet to six hundred ships at a rate equal to or exceeding that planned by President Ford. We will build more aircraft carriers, submarines, and amphibious ships."

Secretary Lehman entered office vowing to revitalize strategic thought in the Navy as well, and to establish firm links between strategy and programs.

Efforts within the navy staff to prepare the Maritime Strategy culminated in briefings by the principals, Admiral Watkins and Secretary Lehman, of the Congress, and subsequently in the U.S. Naval Institute supplement that was described as "the most definitive and authoritative statements of the Maritime Strategy that are available in unclassified form." The Watkins article, accompanied by companion pieces authored by Secretary Lehman and commandant of the Marine Corps, Gen. P. X. Kelley (with Maj. Hugh K. O'Donnell Jr.) is striking not for its unique approach but for its broad, long-term continuity with Navy thinking and for its consistent reflection of Navy strategic culture.

The two Reagan administrations, and the subsequent administration of President George H. W. Bush, continued strong support for a large, offensively oriented, forward-deployed, expeditionary Navy. The years of budgetary stringency under Secretary of Defense McNamara were ended, and the Navy was only then able to return to its deeply set roots.

Critics of the Maritime Strategy

To be sure, the Navy, and the Maritime Strategy, had its strong critics. They argued that the strategy was

- Unnecessarily provocative, dangerous, and escalatory;
- Ineffective in accomplishing war objectives or flatly contradictory to them;
- Too independent, and took inadequate advantage of the complementary capabilities of sister services and of allies;
- Too rigid, pivoting as it did on U.S. maritime superiority and continuing Soviet inferiority at sea;
- Merely justification for a large, expensive Navy and, therefore, not really strategy at all; and
- Contrary, in fact hostile, to the objectives of arms control.

Indeed, in his critique of the Maritime Strategy, university professor Barry Posen argued that attacking Soviet ballistic missile submarines in their protected bastions would cause them to "use them or lose them." He concluded that "we now live in the worst of all possible worlds."[11] The Soviet response to the open publication of the U.S. maritime strategy was similarly strikingly emotional: Valentin Falin, a high-level Kremlin spokesman, branded it "remarkably odious," and asserted: "It is hardly possible to imagine anything worse."[12] Virtually all of the criticism involved the perception that an autonomous, arrogant, block-headed Navy was prepared to take excessive, unnecessary risks for little potential strategic gain.

Yet, by depicting the strategy with the use of the current force inventory of ships and aircraft, the Navy avoided criticism that it was employing forces in the future program that would never be realized. Moreover, the strategy included a section on "uncertainties"—a list of ponderables that could not be known in advance, ones that recognized clearly the age-old dictum that no strategy survives first contact with the enemy in wartime.

On the list were such issues as the problematic use of nuclear weapons. Would they be used? When in the war would they be used—at the onset, or not until a situation was deemed to have become critical? What would be the "ground truth" in which nuclear weapons were either seriously threatened or actually employed; that is, what side was winning, and in which direction was the momentum of the battle? Would nuclear weapons be used on ships at sea even if they had not been used ashore first? Would Soviet vows of never using nuclear weapons first hold in wartime, especially if their forces were stalemated or losing?

Other uncertainties had to do with, for example, the staying power of alliances. Would NATO or the Warsaw Treaty Organization survive the shocks of heavy combat, or would they fracture? Would they fail slowly or precipitately, or would they fight on, perhaps for years? Would warning be adequate to permit deployment to defensible positions—to gain or secure sanctuary—or avoid losing the "battle of the first salvo," as Soviet Admiral Gorshkov labeled it. Should the strategy be aimed at resolving a war a quickly as possible, or to emphasize economy of force counting on the long-term weight of effort of the large, productive economies of the West?

Would technology perform within the "fog of war" as it had in testing? Would surprise provide meaningful leverage to the side that employed it? Surprise can take many forms—in the appearance of a new, highly effective, and perhaps terrifying weapon, in a sudden attack from an unanticipated direction, in the revelation of a shocking new tactic or stratagem, in unexpected violations of the laws and norms of warfare, in the failure of previously sound doctrine under new circumstances, in the demonstrated inadequacy of training methods, and in ways not yet even considered.

For navies, getting most of these uncertainties right was vital, for peacetime inventories of ships and sea-based aircraft are never large enough to sustain significant loss of numbers early in a conflict. Ships take a long time to construct, so they are reconstituted only with difficulty and slowly. This makes it very important to err on the side of caution with regard to all of the uncertainties. It points to a built-in conservatism on the part of navies, which are—with a host of good reasons—loathe to assume large risks early in a conflict.

What is important to comprehend in all of this is that for the most part, neither the correctness of the Maritime Strategy nor of the criticism could be supported by reference to experience or history. The world had—and still has—only scant or no experience to which to point for understanding or guidance on the bulk of the activities emphasized by the Maritime Strategy.

On the matter of risk, Churchill provided the reminder that "it is idle to condemn operations because they involve hazard and uncertainty. All war is hazard. Victory is wrested by running risks."[13] The level of the stakes and risks will inevitably attract criticism of military strategies that appear to some to exceed the bounds of prudence. While criticism has the beneficial side effect of encouraging the clarification of one's thinking, there is a danger that owing to the criticism, well founded or not, the United States will be self-deterred from taking the kind of risks that must be borne in time of war in order to prevail.

Most naval weapons of current vintage have never been used in anger. To offer a single, but vital example, the premier anti-submarine warfare platform of the U.S. Navy—the nuclear-powered attack submarine—has no actual combat experience, none. What analytical methods can be brought to bear in trying to understand the interplay of very serious but truly unexplored areas of warfare, therefore, consist of gaming and simulation, campaign analyses, and field exercises and experiments. These, for the most part, are the exclusive province of the military and not openly accessible to the academic community or to others who write critically about military strategy. This does not indicate that the critics are necessarily wrong, only that they should, in recognition of the point, proceed less pretentiously.[14]

Denouement

To those who are concerned about what they perceive to be an independent, dangerous course being set and followed intransigently by the U.S. Navy, it is important to note that sea control has always been its central mission. During those periods of time when the Navy was clearly preponderant at sea, or when it was on the upswing, it has emphasized both forward offensive operations—attack at the source!—to secure control of the seas and power projection operations against enemy forces or territory. At times when international and budgetary climates were not favorable for sea power, the Navy leadership retrenched. In those latter situations it appears that the Navy has either been forced to adopt a less aggressive strategy, or the leadership decided that in view of the forces available and the prevailing threat the best it would be able to do would be to forego—or at least to de-emphasize—forward offensive operations. What is most striking, in the final analysis, is the similarity between the strategic visions of James Watkins, Arleigh Burke, Forrest Sherman, and Alfred Thayer Mahan—not the differences.

The maritime strategy was conceived, and took its place appropriately, as a *planning*, not as an *operational* document. Clearly and correctly, the Unified Commanders shoulder the responsibility for operational planning. The maritime strategy in contrast, as it stated, presented strategy in its ideal setting: that is, given the forces, given expectations about how those forces will perform, given alliances functioning as they were designed, given political will, given Soviet force actions in accordance with intelligence estimates—in short, given that the myriad uncertainties will turn out not to be wholly unfavorable—the strategy still offered only the *direction* in which U.S. maritime forces were to be guided.

In Chapter 2 the origins of the culture were detailed. The environment and historical precedent were found to be strong factors of influence in the formation and refinement of Navy Strategic Culture. The Maritime Strategy of the 1980s was adopted at a particularly opportune time. The union of many factors at that particular time, complemented by the supportive political and budgetary environment, resulted in a coherent, persuasive, and powerful strategy document.

As a direction-finder for naval officers, as a tool for planners and programmers, as a rationale to be used before the Congress to justify programs, as a public affairs instrument to tell the Navy's story, and as a focusing lens for Navy Strategic Culture, the Maritime Strategy of the 1980s was successful and effective. The Maritime Strategy had as its organizing principle its role in the National Military Strategy to engage the Soviet Union in war and emerge victorious. That was then. Today, the Maritime Strategy is an anachronism and the Navy appears to have been cast back into a world that questions the need for a large, capable, expeditionary Navy. The culture has not changed, however, because the environment and the overall objectives of naval warfare have not changed. One of the hallmarks of Navy Culture is adaptability. What better way to prepare to encounter future challenges?

CHAPTER 8

Retrospective

The strategic culture of the Navy—an expeditionary, systems-based approach to warfighting in forward areas of the world's oceans on mobile platforms, offensively oriented, self-reliant, and adaptable to rapidly changing events—has served the organization and the country well, indeed very well, through the years. On occasion, when the political or budgetary climate was unfavorable, elements of the culture were necessarily pushed to the background, but they were always there as a clear signpost to guide the Navy.

Based on history and experience, and tempered powerfully by the environment, that strategic culture bred confidence in seagoing officers of the Navy. Confidence, that is, in their planning, in their equipment, in their training, and in the crews of their ships and aircraft. One of the tenets of the Maritime Strategy of the 1980s was that the navy of the Soviet Union had no seagoing tradition, no history of naval victories in far-flung venues, no key successes over adversary navies. In fact, close to the opposite was the case, for the Soviet Navy was responsible for the greatest maritime disaster of all time. In its last combat action late in World War II the Soviet submarine S-13 sank the German refugee ship, *Wilhelm Gustloff.* Accounts differ about the number of those who perished in the icy Baltic waters in January 1945, but at least five thousand—and perhaps as many as nine thousand—died, most of whom were women and children. This was hardly a legacy of which to be proud.

It was clearly the intention of Navy strategists of the 1980s to seek to bloody the noses of the operators of the Soviet Navy, and Soviet naval aviation very early in a war in order to make them reluctant even to sortie from their four home port areas, none of which enjoyed unhindered day-to-day access to the open oceans. The strategists of three decades ago appreciated well that fighting forces lacking confidence in themselves tend not to fare well in combat. Expectations of defeat function as a self-fulfilling prophecy. A tradition of excellence in war is neither the accidental product of a single inspired combat performance, nor can it be fabricated by the rewriting of history. The words of John Paul Jones: "I wish to have no connection with any ship that does not sail fast, for I intend to go in harm's way," resounded as clearly in the 1980s as they had two hundred years earlier.

There is an implication, however, that a strategic culture grounded so strongly in past practices renders its practitioners unresponsive to change, especially alterations in strategies employed by adversaries. Defense Secretary McNamara was reported to have once remarked: "Oh, let's stop doing it John Paul Jones' way. Can't you have an original thought for a change?" and one of McNamara's assistants told the Chief of Naval Operations at the time: "I know you military fellows have always been taught to get in there with both feet and get it over with, but this [Vietnam] is a different kind of war."[1]

Alastair Johnston suggests that strategic culture is much more, that it "consists of basic assumptions about the orderliness of the strategic environment, that is, about the role of war in human affairs . . . about the nature of the adversary and the threat it poses . . . about the efficacy of the use of force . . . about the ability to control outcomes and to eliminate threats, and the conditions under which applied force is useful."[2] Preceding chapters have addressed these elements in describing Navy Strategic Culture, emphasizing the overpowering shaping role of the open ocean environment. It should have been amply clear that Navy strategists thought about strategy neither as solitaire nor invariably as how John Paul Jones might have acted, but as a process of how to achieve one's objectives by the adaptive application of military force in a dynamic environment against a calculating adversary. As Heraclitus of Ephesus wrote millennia ago: "No man ever steps in the same river twice, for it's not the same river, and he's not the same man."[3]

The approach described by Navy Strategic Culture has been historically based, but it focused on the current and future security environments and threats from a blue-water point of view. Nevertheless, one might still ask whether the story is complete. Have some vital questions that bear on how the Navy might perform, or how it will confront future issues been left unaddressed? Has the strategic culture become anachronistic? Is it too backward-looking? Is it ossified? Is the

Navy "focused on the rear-view mirror," somewhat oblivious to the changes that have taken place in the strategic environment, while cleaving to antiquated ideas? Are there countercultural issues that will pose severe challenges to the prevailing culture?

The Changing Global Security Environment

The security environment has changed significantly from the 1980s. Unquestionably, the demise of the Soviet Union rendered the Maritime Strategy of the 1980s moot. Unquestionably, the terrorist attacks on the United States of September 11, 2001, starkly exposed a startling and frightening development in which individuals, rather than states, could achieve control of weapons of mass destruction—in that instance fully fueled commercial airliners—and use them as weapons of war to penetrate the sanctuary of states, destroy their property, and kill their citizens on a massive scale. Before that time, weapons of mass destruction were considered the exclusive province of sovereign states which, because they do not commit suicide, can be deterred. If the control of easily transportable weapons of mass destruction—chemical, biological, nuclear, or radiological—devolved to fanatical individuals who cannot be deterred by any means, then the global security situation has been altered drastically. And even though defense against such attacks is very difficult and challenging (attacks can occur potentially at any time, anywhere; it is literally impossible to defend all the time everywhere; yet a single attack can have catastrophic effects), those responsible for the security of the citizens of a state are obliged to commit to a widespread, comprehensive defense in depth. They have no choice unless they are prepared to forfeit the independence and sovereignty of their country. Thus did the devastating and deadly attacks of September 11 on U.S. territory give rise to a U.S. Department of Homeland Security and a change in focus and emphasis in providing for the security of the American people.

Navy strategists view homeland security proper more as a matter of domestic law enforcement than of fighting wars. These are strikingly different matters, ones that have extraordinarily divergent objectives, strategies, weapons, and rules of conduct. The so-called Global War on Terror has in many ways conflated the two, to the consternation and confusion of just about everyone in the world.

Navy Strategic Culture is about the conduct of war; it is definitively not about law enforcement. In the United States, warfare is undertaken incident to an act of Congress or an order of the president, acting in his role as commander in chief of the armed forces, according to the U.S. Constitution. Law enforcement can also take place at the national level, with the Federal Bureau of Investigation providing the means, and it can be effected at other, lesser, levels right down to the community level by local police departments. Waging war, however, lies within the sole province of the president. The objective of war is victory over an opponent by securing strategic goals; the objective of law enforcement, on the other hand, is justice through apprehending and punishing violators and giving safe haven, solace, closure, and comfort to victims.

In law enforcement, a crime is committed and the enforcers say: "Who did this? We must find them and apprehend them." In warfare, an act of war is committed, and the warriors say: "We must kill or capture those responsible." So, when a national leader of a country whose homeland is attacked on a massive scale with weapons of mass destruction interprets this as an act of war, he says: "We must attack the perpetrators of this act, and either kill or capture them." "Dead or alive" is the appropriate sentiment in response to an act of war. If, on the other hand, a violation of the laws of the land is at issue, one says instead: "We must bring the perpetrators of this act to justice." There is a world of difference between these two, but from the time of the suicide truck bombing of the U.S. embassy and Marine barracks in Beirut in 1983, through the murder of U.S. service members in the la Belle discotheque in 1986, the bombing of the World Trade Center in 1993, the 1996 destruction of Khobar Towers, the simultaneous bombing of U.S. embassies in Kenya and Tanzania in 1998, right up to the suicide attack on the USS *Cole* while in the Yemeni port of Aden, attacks on U.S. officials, including its military, were treated as law enforcement issues rather than as issues more properly remedied under the laws of armed conflict. Even the attack on the *Cole* brought forth not military retaliation, but FBI investigators to Yemen. That high officials in the U.S. government used both forms—warfare and law enforcement—in their public statements after the devastating attacks on September 11, 2001, served only to muddle what was a most perplexing situation immediately after the airliner attacks and the near-simultaneous distribution of anthrax powder through the U.S. postal system.

In law enforcement, responsible officials seek to use the *minimum force possible* in resolving the crime; in warfare, one seeks to use the *maximum force permissible* so that the conflict can be won as quickly as possible with the least amount of destruction and carnage.[4] Law enforcement is envisioned as a chronic condition in societies: states will always have crime, and so law enforcement must be performed perpetually. War, in contrast, is viewed as an acute, aberrant condition. Peace, of late referred to as "stability," must be restored as quickly as possible and warfare terminated completely. Now, it should be axiomatic that with such a different perspective and approach to the use of force that rules for using deadly force are very divergent in the fields of law enforcement and warfare and, of course, they are. Yet, politicians, judges, academics, and others—such as the press—often do not separate the two, or deliberately conjoin and confuse them in order to pursue their own particular agendas. The culmination of this line of thinking came in the U.S. Supreme Court's 2008 decision in *Boumediene* to grant the right of habeas corpus, a concept central to domestic U.S. law enforcement, to enemy alien warriors.[5]

It is important to note how potential errors factor into the thinking of law enforcement officers and warfighters. In the former, most essential is to guard against "false positives," that is, capturing and prosecuting an innocent person. In warfare, one is much more concerned about "false negatives"—permitting terrorists and aggressors to be released. So, the emphasis in the former is in favor of protecting the rights of those apprehended so that innocents are not convicted; in the latter, the important thing is not so much that the wrong person is captured, than that a dangerous adversary escape to return to the battlespace. This is the pivotal issue in the detainment of those captured since 9/11: Should the major concern be to prevent false positives, to give captives full rights as prisoners of war, habeas corpus rights similar to U.S. citizens, and access to the U.S. federal court system and its voluminous rules of evidence and safeguards against punishing the innocent; or, should the concern center upon avoiding false negatives and detaining captives for as long as necessary to ensure that they are not able to become a future, perhaps severe and near-term, threat?

Federal law enforcement lies primarily within the responsibility of the Coast Guard and the FBI. The Navy's limited law enforcement role has been generally confined to drug traffic interdiction on the high seas. Never desirous of, nor particularly interested in, such duties—despite its remote similarity to the suppression of piracy—the Navy welcomed the Coast Guard's shouldering of more open-ocean responsibility in the initiation of its "Deepwater" programs.[6] Unquestionably, because the attacks on 9/11 took place on U.S. soil, pressure will be exerted to pull the Navy back from its forward orientation and expeditionary propensities. The Navy's new Maritime Strategy, remarked upon in the next chapter, takes some initial, and vital, steps in this regard. It explicitly states: "Consistent with the National Fleet Policy, Coast Guard forces must be able to operate as part of a joint task force thousands of miles from our shores, and naval forces must be able to respond to operational tasking close to home when necessary to secure our Nation and support civil authorities."[7]

Nevertheless, the serial invasions of Afghanistan and Iraq by U.S. military forces confirmed the U.S. government's understanding that potential attackers are optimally addressed at their source, not in the vicinity of the target—the identical strategic conclusion the Navy came to in the early stages of its existence. The war in Afghanistan, beginning in late 2001 demonstrated for the first time, however, that U.S. military land power could be projected directly, and rapidly, deep into the heart of the Asian continent. Marine, Army, and Special Forces were inserted, and supported, in areas and at distances from support bases that planners traditionally considered nigh impossible. Marines, for example, typically were considered to be configured to fight "in littoral areas." Their reach extended only to on the order of one hundred miles inland from their sea base. With the Afghan incursion, that changed dramatically.

The Marine Corps and "Jointness"

Reference to the Marine Corps raises the question: Why have the Marines been ignored in this rendering of Navy Strategic Culture? Are they not "naval" as well as the surface, aviation, and submarine dark blue–suit community? The answer is straightforward: the Marine Corps has its own, individual, strategic culture. It is separate from, but complementary to, Navy Strategic Culture.

On tactical and logistical matters, the Navy and the Marine Corps, which share the same departmental slice of the defense budget, have their differences. They have had to do primarily with dissimilar approaches to command and control over amphibious operations, and frictions between levels of warfare. The command and control issues have, for the most part, been resolved; the connected issue about levels of warfare extend back to the Guadalcanal campaign in World War II. There is no need to review those here, except to acknowledge that they arose because of the incongruent assessments of Navy and Marine Corps commanders at the operational and strategic levels. These issues extend only rarely to the realm of Strategic Culture.

Some years ago the most accomplished practitioner of amphibious warfare, Adm. Richmond Kelly Turner, put it most succinctly: "We found the most important technique of amphibious warfare to be the willingness and ability to cooperate in spite of differences of opinion or viewpoint between individuals, between branches in each Service and between the different Services themselves, including Allied Services."[8] The implication of Admiral Turner's remark is that the services were in agreement on strategic matters, and willing to work out issues at lower levels. On theme, Frank Hoffman writes, about the Marines: "The most relevant cultural characteristic is what I call their expeditionary ethos. This ethos is the most critical contributor to the Corps' success in combat, especially in the Small Wars and complex contingencies, where the Marines excel."[9]

Similarly, a chronic complaint about the Navy and its Strategic Culture has historically been that the Navy is not "joint" enough: it does not seek to work synergistically with the other services to create a unified fighting force. Instead the Navy is viewed in some quarters as arrogant, independent, and uncooperative.

To some extent, the criticism is accurate. One might point out, however, that the Navy, with its own "air force" (naval aviation) and its own "ground forces" (the Marine Corps), has always been "joint." The Navy knows quite well how to integrate air, land, and sea warfare, for that is what expeditionary warfare is all about. That the Navy has not been overly enthusiastic in working with the Army and the Air Force stems, at least in part, from a belief that for the Navy jointness tends to be viewed essentially as a one-way street. The Navy provides combat services and support services to the other branches of the U.S. armed forces, but receives disproportionately less in return. Of course, the Air Force's aerial tanking and airborne reconnaissance services have been invaluable to the need, under combat conditions prevalent in the Persian Gulf area, to loiter and await the assignment of an air-to-ground target. Naval forces on strike missions historically—take the Korean and Vietnam Wars, for example—have not previously had the need for extended, persistent ground and air reconnaissance and tanking. That need typifies non-linear, counter-insurgency warfare. "In the Gulf War, for example, 20 percent of targets were selected after aircraft launch, whereas over Kosovo 43 percent of targets were selected once the aircraft were airborne. In Afghanistan, 80 percent of the carrier-based sorties were launched without designated targets. Operation Iraqi Freedom statistics are likely to be consistent with this trend."[10] Space-based provision of navigation, surveillance, and communications capabilities are truly national assets, rather than dedicated service-provided ones.

The Army, and to a lesser extent the Air Force, is critically dependent on support from the other services; the Navy can operate for extended periods of time with little or no support from outside the service. The Navy also conducts missions that have virtually no joint participation—anti-submarine warfare, for example. It should not be surprising, therefore, that the Army, and not the Navy, is the primary promoter of jointness. While jointness probably has, or will have, scant effect on forward operations by Navy and Marine Corps forces, it can have an impact on their mobility, offensive power, and self-reliance by reducing their flexibility, and consequently their adaptability. The impact on these core factors by the requirement that operations be joint must be considered in every instance, the point being that jointness does not necessarily equate to unqualified goodness. Insofar as it does not interfere with the Navy's focus: forward, offensive, expeditionary, self-reliant, and adaptive, therefore, the Navy should be more accurately described as indifferent rather than hostile to jointness.

On the other hand, as detailed in Chapter 4, when it comes to cooperation with other navies of the world, the U.S. Navy has no equal. The 2007 *Cooperative Strategy* places strong emphasis on the contributions of navies around the world to a form of global, networked security for all.

Mine and Riverine Warfare

Other areas in which the Navy has been criticized have been its putative lack of interest in, and programmatic attention to, mine and riverine warfare, and the increasing vulnerability of surface warships to "anti-access" forces. On the question of mine warfare, the Navy has long been the champion of offensive mining: placing mines in enemy harbors or across geographic choke points to alter geography to restrict the operations of enemy ships. As early as 1965, for example, the Chief of Naval Operations favored the mining of Haiphong Harbor to prevent logistic resupply of North Vietnam from the sea.[11] In 1972, "Once the minefield in Haiphong was established, Vice Admiral Mack [Commander Seventh Fleet] said: 'What happened was that all traffic into Vietnam, except across the Chinese border stopped. Within ten days, there was not a missile or a shell being fired at us from the beach. The North Vietnamese ran out of ammunition, just as we always said they would.' "[12] Mine countermeasures present a different picture, both in fact and in perception. During the Cold War the Navy cut a deal with its NATO allies, which was, in the most straightforward terms: "We'll bring the tactical aviation to the fight in our carriers, and we'll provide the combatants to protect the carriers and transoceanic surface shipping, we'll cover whatever amphibious operations might be necessary, we'll operate a nuclear deterrent force in the form of ballistic missile submarines (plus bombers and ICBMs, of course), and we'll build a large nuclear-powered attack submarine force. For your part of the naval bargain—you who are much more concerned about a threat posed by the Warsaw Pact land and air forces—you can economize on naval forces by providing the bulk of the mine countermeasures capability for the alliance." This was an attractive bargain for both parties. But it looked for all the uninformed world that the United States was flagrantly neglecting the countermeasures side of mine warfare.

After the demise of the Soviet Union, the Navy surveyed the international scene and noted that the most serious mine warfare hazard would be the threat to stymie a large—say, brigade-size—opposed amphibious operation. Operations that involve employment of the heavy-lift ships of the Maritime Prepositioning

Force do not qualify because, by definition, they take place in a benign, unopposed environment. If there were mines to be overcome, such an operation would not be undertaken.

History counsels that amphibious operations require secure control of the air in the amphibious objective area because landing forces are excessively vulnerable to air attack. Such operations also require extended periods of time to assemble and sail the necessary assault and support shipping. The length of time it would take to gain sufficient control of the air and to organize and deploy a large amphibious task force should be sufficient to relocate minesweeping forces to the amphibious objective area and deal with the mine warfare threat.

As with all the other aspects of maritime warfare, the threat from sea mines is best countered by preventing the mines from being laid in the first place. In keeping with the conviction that the best place to deal with the threat is at the source, interdicting mine laying is the first line of defense. In Operation Desert Storm, for example, restrictions on alliance naval and air operations in the northern Persian Gulf precluded naval commanders from gathering intelligence on Iraqi mining activity, and allowed Iraqi forces to lay mines in international waters and even to set them adrift.

Mine warfare threats to operations of lesser scope—small amphibious raids, for example—could be avoided, or would employ dedicated mine countermeasures on a lesser scale. The effectiveness of mine countermeasures is by no means guaranteed, especially if the mines are planted illegally. Clearly, preventing the laying of mines in the Persian Gulf was precluded for political reasons. As a consequence, mining of international waters of the Persian Gulf by Iraq and the use of free-floating mines by both Iraq and Iran were allowed to proceed, even in flagrant violation of international law. Iraqi bottom mines caused damage to USS *Tripoli* (LPH-10) and USS *Princeton* (CG-59) in 1991, after minesweeping operations in the area had been ongoing for several months. The floating mine laid by an Iranian ship that came close to sinking the USS *Samuel B. Roberts* (FFG-58) three years earlier was struck in international waters transited by that frigate only a few days previously. In summary, mine countermeasure efforts had either been in progress, or would have likely been ineffective against the three successful mining attacks on U.S. combatant ships. New U.S. combatant ships are being designed with deployable mine countermeasures as an optional integral capability in order to mitigate the future mining threat.[13]

Riverine warfare reappeared in Iraq as a Navy mission area, after a hiatus in the U.S. Navy of over thirty years. Patrol boats operated by Navy personnel plied the waters of the Euphrates River and Lake Qadisiya, as the Navy stepped forward to relieve the Marines of this task. Riverine operations, undertaken also during the Vietnam War by the Navy, represent a small part of the "expeditionary" aspect of the culture. The Navy Expeditionary Combat Command—stood up in early 2006—includes in its capability package not only riverine warfare, but explosive ordnance disposal; maritime expeditionary security; diving operations; naval construction; maritime civil affairs; combat camera, and expeditionary training, logistics, medical, combat readiness, and intelligence.[14]

Still, it should be noted that virtually all functions of naval forces are conducted on, over, under, or from the seas. Riverine forces are not, strictly speaking, naval forces. They are essentially land forces adapted to operate on rivers, and perhaps in estuary areas, but they do not depart from territorial waters. Being different from naval forces, they are uncommon for navies, and the strategic culture that accompanies naval forces is unsuited to and unsuitable for riverine operations. Special operations on rivers are not integrated with naval forces because they do not comprise operations on, over, under, or from seaward axes. As an anomaly for navies, they play no part in the formation, activities, or concepts that shape Navy Strategic Culture.

Navy forces are not extensively trained in riverine warfare. Some tactics can be resurrected from the Vietnam era, but riverine warfare is not destined to become the centerpiece, or even an important slice, of the Navy's expeditionary efforts. "The Navy is not primarily about low-level raiding, piracy patrols, and riverine warfare," said Jim Thomas, a former deputy assistant secretary of defense. "If we delude ourselves into thinking that it is," he continued, "we're finished as a great power."[15]

Anti-Access in the Twenty-First Century

The high cost of warships and the striking power of modern missiles, offering the opportunity of sinking a billion-dollar ship with a missile costing hundreds, or even tens of thousands of dollars, has ignited analytical interest in the United States. The issue is whether, in an era of long-range weapons and improved reconnaissance and surveillance, surface ships can be effective in conducting operations in forward areas—that is, near to the coasts of a capable adversary—or are the unfavorable cost-effect ratios and the potential loss of life in large numbers not worth the potential gains? Much analytical effort and thought has been devoted to this question, which, like most such endeavors, depends grandly on one's assumptions about the interaction of opposing forces.[16]

It is time to recall that the most difficult problem in naval warfare is finding the adversary. Finding ships at sea requires over-the-horizon sensing. That means the employment of overhead detectors of some type—satellites, aircraft, drones, aerostats, or blimps. It also means command and control capabilities to evaluate information, and then to prepare and direct a response that is time-sensitive, for the intended targets are all mobile. It also involves sophisticated long-range attack capability, either aircraft or missiles. Two things are clear: first, the investment must be considerable for a state to design and assemble an anti-access system that would be effective against modern surface combatant ships; and second, the intended adversary for anti-access efforts is the U.S. Navy, for no other country possesses a Navy with large expeditionary capability. To prevent access by submarines—either attaining a submerged launch position for missile attack, or perhaps to land a small clandestine group of commandoes—would involve orders of magnitude greater difficulty and investment.

The problem is not as straightforward as it appears. Anti-access capabilities are tightly interlinked. The component elements of an anti-access force are intelligence, reconnaissance, surveillance, command and control, and weapons capable of homing on ships and overcoming their defensive systems. The anti-access chain can be attacked at any point. One could work on breaking any individual part, for success against any component would cause the system to fail.

If intelligence is denied to the adversary, or deception is successful on the part of approaching forces, anti-access will fail. If reconnaissance or surveillance forces are thwarted, anti-access will fail. If command and control is rendered ineffective by countermeasures, anti-access will fail. And if the weapons can be defeated en route to their targets, anti-access will fail. You can't hit what you don't know is there. You can't hit what you can't find. You can't hit what you can't track. And you can't hit something that has an effective defense against your weapons. So, it is far from obvious that even a transparent battlespace and flawlessly performing weapons will make access unacceptably risky in the future. Finally, it is the hallmark of great navies throughout history that they are willing to run the risks of losing ships in order to accomplish their mission. States with no history and tradition of going forward aggressively and conducting military action would find themselves cowed by anti-access forces. The U.S. military, on the other hand, in contemplating the employment of naval forces, respects anti-access capability, but cannot be stymied by it.

Civil-Military Relations

Changes in the *national* strategic culture, which manifest themselves in civil-military relations, are likely to have a transformational impact on Navy Strategic Culture. As Colin Gray has observed, "Revolutionary change in warfare may be less important than revolutionary change in social attitudes to war and the military."[17] A comprehensive analysis of the impact of future civil-military relations would be inappropriate here, but some near-horizon issues weigh heavily on Navy Strategic Culture. Two stand out: constraints on the use of military force, and the impact of sexual issues on the Navy.

Constraints on the use of military force arise essentially from the Judeo-Christian heritage of the United States, based in the conviction that disagreements should be worked out peacefully, and that a resort to violence signals a failure to find and implement a compromise solution. The use of military force must be as a last resort, and specific criteria must be met for participation in (the legal doctrine of *jus ad bellum*) and conduct of (*jus in bello*) armed conflict for it to be moral and just. Michael Walzer's *Just and Unjust Wars* provides a comprehensive, nuanced view of the subject.

Constraints in the Western way of war flow from a variety of sources. They can be operational, organizational, legal, and moral in nature. Over time, the West in general, and the United States in particular, has levied and layered constraints on their armed forces, which have been increasingly greater in number and severity. The ubiquity of the twenty-first-century media, and the ability through overhead surveillance and the connectivity of the internet to observe and report on activity virtually anywhere in the world have penetrated the fog of war, to be sure, but more from the point of view of the observer than the warrior.

In the modern battlespace, as in times past, combatants must be prepared to act on the basis of incomplete—and sometimes, regrettably, even incorrect—information. The stakes in the employment of military force are life and death for warriors who operate at the tip of the spear. That they never have full knowledge about the situation or their adversary, and that they often must act either in self-defense or to prevent an adversarial action describe conditions of warfare that have always existed and will always exist. As noted, the Western warrior is confined to the employment of the maximum force permissible, yet that must be determined by the warrior on the scene, in the most stressful of circumstances, and in the absence of perfect knowledge. Accordingly, mistakes will be inevitable, and errors will be both unforced and forced by the adversary, who endeavors to ensure that the context is as confusing and terrifying as possible.

Why this matters is that in order to succeed, Navy Strategic Culture, as it has been discussed herein, must be aggressive in its application. Clearly, in war, "the exercise of maximum violence for swift results has been the American way,"[18] and the Navy has been, and intends to be, no exception. Given the fog and friction of combat, the greater the aggressiveness, the more likely mistakes will be made. Yet, the contemporary strategic environment tends in many complicated ways to render the application of military force more difficult, while at the same time being as intolerant of error as it is impatient for results. This is not only true, but most unfortunate, for it has long been recognized that "the most that can be claimed for a knowledge of the science of war is that it so prepares the student that he may make the least number of mistakes."[19]

Fine-tuning the rules of engagement; overweening concern about collateral damage—including prospective harm to the environment; application of the "pottery-barn" concept of "you break it, you own it";[20] detailed reports on combat operations submitted by unqualified, unverified, or deliberately deceptive sources; intra-war show trials of combatants based on flimsy or trumped-up charges; the tampering in combat operations by a media that is often partisan, and even hostile to operational goals; the intervention of lawyers—both civilian and military—at every stage and at every level of combat operations and prisoner affairs: these are just a few emergent characteristics of twenty-first-century warfare.

The vectors of all of these constraints point to a hampering of the aggressiveness necessary to pursue combat operations against an adversary affected and restrained by none of them. In fact, adversaries benefit directly from them. That is what is truly asymmetrical about twenty-first-century warfare: adversaries take advantage of their opponent's constraints. In fact, current adversaries consider the constraints under which the fighting forces of the West—and particularly those of the United States—labor to be signs of weakness. Superior U.S. technology is rendered unusable by adversaries who flout not only the laws of armed conflict, but the most fundamental of natural laws—the deliberate murder of innocents. When they employ children and the mentally infirm as sacrificial weapons, have adversaries not reached the bottom rung on the ladder of immorality? What, after all, are IEDs (Improvised Explosive Devices) other than land mines, the use of which has been banned by international convention specifically because they can be indiscriminate killers of the innocent? When adversaries seek weapons of mass destruction, exhibiting every intention to slaughter large numbers of noncom-batants, how can they be deterred?

Aggressors, and potential aggressors—yesterday the Nazis, the Japanese militarists, the fascist Italians, the communist Soviets, today and tomorrow the Islamist fanatics and contemporary totalitarian dictators—know that successful strategies attack their adversary's weaknesses. The constraints on Western forces that make the situation so markedly asymmetrical tilt the battlespace in ways that are extremely difficult to counterbalance.

One wonders whether tomorrow when the captain of the U.S. aircraft carrier launches a strike if the pilots will question in detail what their targets will be and whether they have on board their aircraft the proper munitions with which to attack them—cluster munitions, for example, or bombs with a blast radius too extreme for the target. Will the pilot question the accuracy and the reliability of the Marine on the ground who requests suppressing fire on a particular house or position? Will the crew of a gun mount query the captain of the cruiser eight miles at sea on the appropriateness of the target they are engaging, which they (and he) cannot see and cannot evaluate? Will they request a lawyer's advice before pulling the trigger? The madness of the search for an antiseptic, thoroughly legalistic, error-free way of war threatens to permeate all levels, to increase the risks of failure, and to offset significantly the advantages offered by superior technology, training, leadership, and the application of proven successful strategic culture.

What makes the military calling unique is a willingness to accept the risk of being called upon to make the ultimate sacrifice. Credibility, both with one's own decision makers and with adversaries and prospective adversaries, is vital to the success of military forces. U.S. officers take an oath to perform their obligations "well and faithfully." This cannot be taken for granted, but must be *indicated* in some way. Yet, *indications* of military capability cannot rely on actual combat operations, for there might have been no opportunity or occasion to undertake them. They must be demonstrated in other ways—shows of force, experiments, and exercises, for example.

To the extent that confusion reigns about the ability, the latitude, or the willingness to use force, both friends and adversaries wind up perplexed. Is there any doubt that Saddam Hussein was surprised and shocked by the counterattack in Kuwait and Iraq by the U.S.-led coalition in 1991? Is it even imaginable that Osama bin Laden was not surprised at the intensity of U.S. operations against Afghanistan and then against Iraq? After all, he had been conditioned by a series of feeble responses to attacks on Americans—including the American military directly—to believe that the United States was the "weak horse," as he claimed.[21] Was not the U.S. response to the attack on the USS *Cole* in Aden to send the FBI to Yemen in order to apprehend the criminals?

The Army and the Marine Corps have been most significantly affected by a pervasive reaching into the battlespace and altering the conditions under which they conduct their combat operations. Because these efforts to do so have come not only from the political process in the United States, but from friends, allies, and enemies alike; from non-governmental organizations; and from an adversarial press, domestic and international, the impact cannot be confined only to the ground forces. It is felt across all the U.S. armed forces, and it serves to compromise their technological advantages—acquired at significant expense—their unexcelled training, their top-notch leadership, and their fighting spirit. The effects are troublesome and corrosive, for they lead inexorably to significantly increased risks and to the potential defeat of the most effective, efficient, and humane fighting force the world has ever seen: defeat, not at the hands of the enemy, but rather because of the cumulative weight of the constraints on their ability to achieve the missions they have been given.

Throughout the Iraq War beginning in 2003 constraints have continued to build. The adversary has invariably been given a pass for its actions, most of which have been contrary to the laws of armed conflict and even to basic humanitarian law, even while U.S. forces were condemned for "harsh" treatment of prisoners, excessive collateral damage, insufficient attention to the protection of noncombatants, and failure to provide detainees with basic needs and judicial protection. The political polarization is so great that efforts to explain that wars are bloody affairs and that the instruments of war tend to be blunt because the information on which they are applied is invariably imperfect fall only on deaf ears. Even efforts to recruit for the all-volunteer military force have met with organized resistance in parts of the country. Remedies for this set of severe problems are not evident; consequently, it is easy to be pessimistic about the future ability of a volunteer U.S. military to keep the citizens of the country safe from devastating attack. The complexity and the difficulties imposed by the scope and scale of constraints on the use of force are compounded by the addition of sex as a factor in military service. Chapter 2 examined the cohesive forces of Strategic Culture, and suggested that teamwork, cemented by the bonding of groups of equals, was of central importance to prevailing over the environment and the adversary. Chapter 3 investigated the context in which naval operations are conducted, and illuminated the hazardous nature of seafaring and conducting warfare on, over, and from the sea. It highlighted the fact that the environment in which navies operate is far removed from that on land, not only geographically but emotionally.

At sea—except for what's in the skies and the heavens—not much is familiar. Beyond the sight of land, everything is different and everything is unorganized. Whether one is a surface warfare officer, an aviator, or a submariner, one's ship becomes home, both physically and psychologically. The organization of ships is hierarchical, and it revolves around systems. At sea, all of the body's senses and appetites are required to respond to an environment filled with unfamiliar situations and demands, ones that persist for fully twenty-four hours every day.

A graphic circulating on the internet illustrates how a hag morphs in successive stages into a beauty, which takes place coincidentally with the number of glasses of beer one virtually consumes. As time away from home and time away from familiar things increases, so does the sailor's longing for them. The caricature of the nineteen-year-old seaman, who had never before been away from home for any significant time, hitting port and "making liberty" after a six-week stint at sea, his hormones raging, with a wad of money in his pocket (which he had scant opportunity to spend at sea), an unslakeable thirst, and a hankering for some female companionship that looks more and more desirable the longer he has been outside the sight of land is more platitude than parody.

Naval officers understand their ships and aircraft as systems within systems, and they understand how systems depend upon the effective and efficient interworking of all their component parts. Commanders of naval forces are vitally concerned about the readiness of their forces: readiness to weather the elements; and, simultaneously, readiness to act as ambassadors of peace, to demonstrate resolve by a show of force, or to transition into full combat operations literally at a moment's notice. This unparalleled ability to make a seamless conversion from placid cruising on the open seas to full-up combat is a hallmark of the naval service, a pillar of Navy Strategic Culture, and an unmistakable measure of its overall contribution to the security of the nation. "Only 15 percent of some 536,000 troops in Vietnam were combat soldiers," wrote Victor Davis Hanson in *Carnage and Culture*.[22] Everyone on a combatant Navy ship is a "combat soldier."

According to law, Title 10, U.S. Code: "The Navy shall be organized, trained, and equipped primarily for prompt and sustained combat incident to operations at sea." Throughout its existence, the Navy has held firm to the understanding that this could best be accomplished by a well-trained systems-oriented force that was expeditionary: forward, offensive, self-reliant, and adaptive.

Sailors who spend extended periods of time at sea find their thinking altered by their environment, even as they bind together in groups. They are bound by the relentless nature of the physical environment and the potential for combat operations, both of which touch upon their deepest human instinct—survival. Training to attain excellence is continual and pervasive. Over time, sailors come to understand that the culture sanctions, and results of combat reward, operations that are offensive and require the assumption of risk.

Underwriting the readiness of their ships, sailors bond together, hopeful always that they will not be called upon to unleash the violence for which they have been trained, but determined to prevail should the demand arise. Group cohesion and teamwork is paramount because the stakes and the risks are high and the margin for error can be slim.

Male bonding—a demonstrated effective method of implementing key elements of the culture—has a storied history in military organizations, and particularly in navies, which, physically isolated from civil society for long periods of time, leads to the formation of close-knit single-sex communities. To build trust and loyalty, military organizations employ discipline, and simultaneously promote a spontaneous willingness to follow orders to perform difficult and perhaps the most unpleasant acts. Teamwork is fostered and nourished to focus energy and effort in order to fulfill mission requirements. Teamwork is grounded in the *equality* of members of the team. It typifies Horatio Nelson's "band of brothers" approach to naval warfare, and characterizes the most professional fighting units throughout history. Brothers form teams of equals, and join together to attain their objectives. Shared heroes and histories, as well as the sharing of a common fate, foster emulation and bonding.

Bonding assumes several forms. As Alastair Finlan has written, "Close relationships between male and female officers (and other ranks) in confined spaces like warships . . . [in an environment] that celebrates hypersexuality inevitably creates bonding at the sexual level."[23] This is little more than common sense. Indeed, "There is not a sailor alive in any navy who has not occasionally wondered what on earth he is doing flogging the ocean for months on end when he could be at home with his family. The introduction in ships of more women, some of them mothers, can only serve to reinforce that feeling," wrote Richard Sharpe.[24] For a ship's captain, the substitution of fifty women for men in a crew of four hundred introduces a new, unwelcome source of tension within the ship and creates a virtual infinite number of love triangles.

The natures of men and women, underwritten by varied genetic and hormonal substructures, are radically different. This means that *inequality* is innate, both in fact and perceptually. Yet, U.S. society has "adopted, quite without realizing the magnitude of the change, a practice of equality between the sexes that has never been known before in all human history."[25] The "practice of equality" Mansfield refers to is "equality under the law," not the equality that men must understand in order to bond together in a spiritual quality that induces "manly men to risk their lives in order to save their lives."[26]

"The term 'cohesion' describes the bonding of a unit's fighters that converts it from a group into a force," wrote Wayne Hughes. And the link to Navy Strategic Culture, grounded in a systems approach is evident, as he continued, "A cohesive force functions as a unified, interconnected system."[27]

The *displacement* of males by females in this environment, never before even attempted, creates the potential to disrupt the interworking of the system, reduce combat readiness significantly and, accordingly, markedly increase the risks attendant to combat. As the Presidential Commission on the Assignment of Women in the Armed Forces observed, "Warfare is a supranational survival contest in which opposing sides vie for any advantage; unilateral policies adopted to promote principles other than military necessity may place the adopting party at increased risk of defeat."[28] This strikes at the very taproot of Navy Strategic Culture—its boldness.

Recall that U.S. naval ships have historically been ascetic settings. Distractions—vivid reminders of life ashore—have been not only minimal, but also minimized. Women traditionally have been banned from ships at sea, as have alcoholic beverages (since 1917) and illicit drugs. The absence of such diversions encouraged the crews of ships and aircraft to concentrate on the business at hand: mastering the elements, operating their systems, and preparing for combat. Entertainment outside of working hours and while not on watch consisted of writing letters, engaging in bull sessions, playing cards or acey-ducey, reading, and the nightly "flick" on the mess deck (dining hall).

In the past thirty years or so, the number and quality of diversions have increased manyfold. Starting with the portable individual cassette tape player, now primitive, entertainment systems have evolved to the point where today sailors can have their personal first-run movie theater on their lap, wherever they are, for about a hundred dollar investment. Hand-held electronic games, personal digital assistants, cellular telephones, and internet access devices have become omnipresent. Illegal drugs are portable, available, and difficult to detect. All of these compete for the attention of crewmembers, and divert their attention from the business of operating ships safely and preparing for combat. They all inhibit the bonding of the crew and building the professionalism that is necessary to achieve high levels of readiness. Nevertheless, adjustments and accommodations have been made, and ship captains still marvel at how remarkably readiness increases when a ship takes to sea and leaves the bulk of the distractions and temptations of landsmen behind.

The introduction of women into the crews of ships brought a wrinkle with unprecedented physical and psychological effects. For men, the sight of a woman, any woman, resurrects the most primal of emotions: of mother, of heterosexual friendship, of love, of nurturing foregone, and of sex. The ancient proverb "Men love with their eyes; women, with their ears," is nowhere more clear. The absence of women onboard means that the men must revert to virtual images to resurrect these remembrances; the presence of women, on the other hand, is a living, constant reminder of their separation from these vital needs, and of their personal isolation from them.

This, sex, is the central issue with regard to women on Navy ships. The issue is not one of "equal rights," for there is no right to serve in the armed forces of the United States. The armed forces are not an equal opportunity employer. They discriminate, for example, on the basis of physical size and health of potential service members, and on citizenship status, age, educational level, criminal record (called "good moral standing"), and aptitude for service. It has nothing whatever to do with whether women are physically or psychologically capable of performing satisfactorily onboard ship, or any of the myriad of other reasons that have been suggested for their inclusion. That women have served aboard ship successfully and honorably is simply irrelevant. In the final analysis, it is the effect their very presence has on the functioning of the system and its ability to meet its requirements that matters.

Differences between men and women are fundamental and innate; they reside in their hormones. Citing a study in the *Annual Review of Psychology*, Kingsley Browne writes: "Androgens produce a male mind oriented more toward risk, aggression, competition, and dominance, just as they produce a male anatomy that is larger, hairier, and more muscular."[29] Key divergences are not in attitudes or capabilities; they are in the fundamental physiological building blocks of the human body. Furthermore, they reach down into primal instincts as well: "The psychological predisposition for men to prefer masculine comrades in arms and to resist inclusion of women in warfare evolved because it was in their survival interest—and therefore their reproductive interest—to do so. Similarly, women evolved risk-averse tendencies because it was in *their* reproductive interest to do so."[30] It is noteworthy that women have been excluded from ground combat assignments, but are permitted on combatant ships, even though one of the pivotal strengths of the Navy is its ability—in the exercise of its strategic culture—to transition quickly from peacetime to full combat posture.

Survival and sex are fundamental biological drives. With regard to the bonding of warriors, so vital to unit cohesion, survival and sex pull in opposite directions. "The basis of solidarity in any group," writes Lee Harris, "Is the shaming code that has been instilled in all its members."[31] Calls to "get over it" seek to deal with core biological issues at the cognitive level. Or, as Mr. Harris continues, "You cannot argue people out of their shaming code, nor change it by an appeal to logic or empirical evidence. The visceral shaming code resists all efforts to repeal it."[32] One might be amused at the ingenuousness of those who do not grant the issue the weight it deserves, but the way it has been treated mirrors the level of discussion that accompanies many social issues in the United States today.

Homosexuals represent yet another facet of the same issue: the introduction of sexual tension into an environment where it was either previously absent or strongly suppressed. Sex, either homosexual or heterosexual, exposes deep conflicts within the human psyche and evokes the strongest of human emotions: resentment, jealousy, lust, envy, favoritism, and even hatred. All of these run counter to the forces that bind warriors together.

Nearly a decade ago a photograph circulating in email showed a NATO navy frigate conducting an alongside transfer. Members of the crew were standing at the lifelines, and a sign in several languages was displayed reading "I hope we shall meet again." Six of the female crewmembers facing the camera have exposed their breasts. Reflected against the discussion earlier about the need for seriousness of military forces, in the absence of actual combat to be indicated, it bears a message similar to that of the photographs of prisoner abuse at Abu Ghraib. Regardless of how one interprets the photograph, it will stand out permanently as a reflection of a lack of professionalism and weakness–not the confident image of strength that a warship should project.

One searches in vain for a claim, made authoritatively and responsibly, that the introduction of women, and prospectively open homosexuals, to shipboard life has, or will be, beneficial to meeting the Navy's military requirements, which are established by law. The reasons offered for inclusion of women and homosexuals center instead on "equality," or "the desire to serve," or "equal opportunity." But the result is to introduce a factor, a fundamental human factor, into an environment where it did not exist before, creating situations that can only be described as most deleterious. For example, the U.S. Department of Defense Gender Relations survey, conducted by the Defense Manpower Data Center (DMDC) in 2007, found that 34 percent of active duty women and 6 percent of active duty men indicated experiencing sexual harassment. This represents a chronic problem, no solution for which can be foreseen.

It should come as no surprise that naval commanders are hostile to anything that imperils their readiness to perform competently at all levels. The Strategic Culture will necessarily be forced to change to accommodate the decrease in group cohesion that will result. Some kind of softer, gentler, feminized culture will replace the aggressive one that has fostered a legacy unexcelled throughout the annals of sea power. On less majestic levels, there will be pressure to increase crew size because of the incidence of pregnancy and the higher rate of female incapacitation, to deal with privacy in berthing and toileting for both men and women, and to facilitate the transfer of personnel among work groups to alleviate sexual tensions within groups. These are only a few of the less important accommodations that will need to be effected. If the presence of women on combatant ships—including, of course, in naval aviation—is allowed to continue, and perhaps expand into submarines, and if homosexuals are permitted to serve openly in the armed forces, then the potential for catastrophic failure in future combat will have been increased.

The risks and stakes for the United States of defeat in armed combat are the highest imaginable. Today, the devolution of control of weapons of mass destruction to the personal from the state level poses a potent existential threat to the United States. There is no laboratory, no exercise, to test whether the resultant reduction in combat effectiveness by the introduction of women, and prospectively open homosexuals, into combat operations will result in the loss of a war, but those are precisely the stakes.

Once again, as Lee Harris pungently observed: "Every culture has both core values and incidental ones. The core values cannot be challenged; the incidental ones may be . . . excluding females is a core value. It is a tradition that cannot be altered without bringing the whole edifice of the culture tumbling down."[33]

While the risks and stakes are the highest imaginable, there are those who fight for inclusion of women and open homosexuals in the combat arms of the U.S. military, something entirely novel in U.S. history. They tinker with such matters, not having any idea of the consequences they might wreak, but there is no way short of losing a war that would demonstrate how devastating such a policy could be. It is certain as certain can be, however, that such advocates will shun all responsibility if the United States should subsequently lose a war. Failure will not be because of their insistence of inclusion of women and homosexuals into the military, contrary to history, logic, and common sense. Instead, the blame will fall on flawed planning, deficient leadership, incomplete or incorrect intelligence, incompetent execution, ineffective or improper weaponry, or a multitude of other reasons. So, for the advocates, it is—foreseeably and regrettably—an *experiment* totally free of *accountability*.

A knowledgeable observer has written: "The Navy, even more than the other services, has a reputation for resistance to change. Given the continuous and dramatic changes absorbed, the reputation is undeserved. Perhaps the defiant nature of the institution resists instead change originating from outside the sea service. The Navy has been as separated from the public as the Army has been connected to it."[34] This separation makes it difficult for those who have never been beyond the sight of land to understand the culture of the mariner, and especially those who, as a profession, participate in the exercise of sea power.

CHAPTER 9

Conclusion

The night was dark, very dark. It was the time of the new moon, and clouds obscured the stars. Darkened, so that no light was visible outside its skin, the ship glided almost noiselessly through the glassy sea. In the near absence of light and sound, the sense of isolation from everything else in the universe was all but complete. In wartime, against an enemy capable of striking at any time from any direction without warning, the tension level would be high and unrelenting. In peacetime, with no palpable threat, it would offer prime time for reflection.

With the sunrise, a glorious day dawned, but the seas were building and the wind freshening. The barometer was falling, and the meteorologists predicted a sharp, nasty storm lying directly on the ship's track. Watertight doors were dogged down, and hatches battened; nothing was left adrift, for it was going to be a rough transit.

In an often quoted passage, Samuel Johnson wrote, "No man will be a sailor who has the contrivance to get himself into jail, for being in a ship is being in a jail, with the chance of being drowned. . . . A man in jail has more room, better food, and commonly better company."[1] Old Dr. Johnson was keenly aware of the environment, and hence the power of the watery context.

At sea, context rules. The weather, the seas, the isolation, the unfamiliarity of the environment, the political vacuum, the scarcity of fresh water, the absence of landmarks, the low level of traffic, the lack of lines in the form of roads and boundaries to organize and direct activity—all weigh upon and condition the thinking of mariners. If those mariners are warriors, they know they must first master the oceans with their special navigational and weather systems. Only once that has been accomplished can they turn their attention to other important matters.

Given powerful oceangoing ships and their integral systems, and given weapons to strike powerful blows at long ranges on adversaries in the air, on the surface of the planet, or beneath the seas, seagoing warriors adapt to their environment and adopt an aggressive approach to fighting. At least, that is what happened in the U.S. Navy. Over time, driven by circumstance and supported by stout ships, first-class systems and weapons, and loyal and committed sailors, the U.S. Navy devised and developed a strategic culture that was expeditionary: forward, mobile, offensive, self-reliant, and adaptable. The U.S. Navy was forward "from the start"[2] launching a small Marine Corps expeditionary operation against a British fort at Nassau, Bahamas in 1776, and then appreciably larger and prolonged operations against the North African coastal states just after the turn of the eighteenth century.

The expeditionary tradition continued throughout the U.S. Navy's history, reaching its culmination in the "island-hopping" campaigns in the Pacific against the Japanese empire from 1942 to 1945. Almost immediately after World War II the United States began deploying its aircraft carriers to patrol the Mediterranean Sea, and when the carrier *Franklin D. Roosevelt* sailed into the Mediterranean in 1946 the 123 onboard aircraft outnumbered the aerial combat orders of battle of any Mediterranean littoral state. Subsequently, the strategy of the North Atlantic Treaty Organization (NATO) specifically called for the continuous deployment of U.S. carrier battlegroups to the NATO European maritime areas of operation; and, in the Pacific, Japan offered a home port to a U.S. aircraft carrier.

The U.S. Navy's Maritime Strategy of the latter part of the twentieth century, which focused narrowly on a global war against the Soviet Union and its allies, has been overtaken by events and superseded. With varying names and emphases, a series of documents has subsequently been issued from the Pentagon about Naval Strategy.[3] In the main, these have not qualified as operational, or even planning strategies, for while they do address the "ways" naval forces are to be employed, they are unspecific as to where, when, and against whom maritime combat power is to be applied. Strictly speaking, they should be considered concepts to guide the employment, acquisition, and development of naval forces.

Issued in the autumn of 2007, the Navy's *Cooperative Strategy for 21st Century Seapower* also should be considered as a conceptual document rather than a strategy, for the same reasons. It asserts: "Where conflict threatens the global system and our national interests, maritime forces will be ready to respond alongside other elements of national and multi-national power, to give political leaders a range of options for deterrence, escalation and de-escalation."[4] And, it seeks to build upon what J. C. Wylie wrote over four decades ago: "The maritime strategies are the one field in which the United States has an inherent advantage over any enemy. The sailor hopes the nation, if it is ever forced to war, will take advantage of that, use it, and exploit it for all it is worth."[5]

The *Cooperative Strategy* does retain and perpetuate the elemental parts of the Navy Strategic Culture, while emphasizing increased integration and interoperability among U.S. naval forces and the navies of most of the states of the world for mutual security and political stability, and with the U.S. Coast Guard for U.S. homeland defense efforts. It seeks to describe what is foremost in the minds of Navy strategists: how naval forces can make a strategic difference.

One might appropriately ask at this juncture what benefit the Navy seeks from the continuation of its approach to the employment of naval forces in peace, crisis, and war. After all, Thomas Friedman contends that the global context has changed dramatically. Indeed, he argues that *The World Is Flat*.[6] Friedman suggests that for a host of reasons the economies, polities, environments, and cultures of the world have become highly integrated and interdependent. Increases in the widespread access to and ease of use of computers, high-speed connective circuitry affording access to information—unimaginable just a few years ago—and the changes these have wrought in business, government, and cultural practices have served both to shrink and to flatten the world.

Globalization, by facilitating the movement of people, ideas, and finances across great distances and through national boundaries, empowers people in many aspects of global activity. To individuals it offers opportunities previously enjoyed only by sovereign states, and that includes the ability to employ deadly force in massive ways. And it reaches into cultures in ways heretofore neither considered, nor even thought possible. Victor Davis Hanson has written in this regard: "The deep-seated anger and humiliation of Al Qaeda were . . . incited by a globalized and Western culture that really did threaten all the old hierarchies of an increasingly dysfunctional Arab and Islamic world."[7]

Much of Mr. Friedman's Flat World moves at the speed of light (162,000 nautical miles per second) and coupled with the low cost of information flows and storage certainly gives it the appearance of being flat. On the other hand, the U.S. Navy knows that the world of large bulk cargoes—especially the over four thousand tankers that carry petroleum, the transfer of which takes place 27 million times slower, on the order of .007 nautical miles per second—is unquestionably round. Information moves at the speed of light, while valuables move around the world by air: people and small items of value that render them worthy of the shorter transit times but higher costs of air shipment.

International regimes that ensure "access to markets, freedom of trade across international boundaries, an orderly state system that prefers peace to war, speedy communications and travel across open seas and skies"[8] promote globalization. The high seas remain politically uncontrolled, but they are free for all to use safely. That freedom and safety is underwritten primarily today and foreseeably by the U.S. Navy, a fact that has long been—and continues to be—taken for granted. The economic well-being of the United States has become more and more dependent upon seaborne commerce, but with globalization so has that of the rest of the world. Indeed, the *Cooperative Strategy* asserts: "The maritime domain—the world's oceans, seas, bays, estuaries, islands, coastal areas, littorals, and the airspace above them—supports 90 percent of the world's trade, it carries the lifeblood of a global system that links every country on earth." Clearly, international economic growth and prosperity have become dependent on globalization, which in turn is dependent upon the unfettered use of the global "commons"—the seas and the air and outer space above them.

In 1959, at a time when the U.S. fleet numbered 860 ships, the Chief of Naval Operations lamented that "we have a 4-Ocean Fleet, but a 7-Ocean Job."[9] Yet, now as then, one of the major reasons that the articulation of a strategy appears as a challenge for the U.S. Navy is that, the protestations of Chiefs of Naval Operations through the years notwithstanding, the U.S. fleet is so commandingly superior. Overwhelming superiority of forces tends to devalue strategy in the maritime realm. With more large aircraft carriers, more small aircraft carriers, more large amphibious ships, and more cruisers in its naval order of battle than the rest of the world combined, the Navy enjoys an embarrassment of riches. But it has these by default—not because it mindlessly sought primacy or empire with no understanding of what to do with them.

In peacetime, the U.S. Navy is deployed around the world to demonstrate its commitment to friends and allies, to place adversaries and potential adversaries on notice, and to underwrite the free use of the seas. Alone among global naval forces, the Navy under its Freedom of Navigation Program seeks to uphold international law on a global scale and to prevent illegitimate claims and practices from becoming accepted as customary.

In the transitional period between peace and war, the Navy intends—as it has historically—to be at the scene, not in far-removed garrisons or bases. If major war should eventuate, the "commons" of the open oceans and globalization will help ensure that it will be global in scope. And, "Every global war has been, decisively, a naval war. Global wars have a naval and maritime complexion because the global system depends for its organisation principally upon intercontinental interactions."[10]

The non-linearity of the maritime battlespace has now migrated over land, and realization of that fact and its ramifications is just beginning to take hold. As Robert Kaplan writes: "The entire earth was now a battlespace. Killing the enemy was easy; it was finding him that was difficult, whether he was concealed amidst civilians on a crowded bazaar street, or lurking in oceanic layers where sound waves traveled and refracted at unpredictable speeds and angles."[11] As in naval warfare since its inception, now the key to land warfare is finding the enemy, and penetrating his sanctuary while simultaneously securing one's own safe haven.

Even as the world has in some ways become flatter and battlespace has become non-linear, the operational context facing U.S. military forces has been altered dramatically in other ways. Today, the United States requires its military forces to:

- Deploy somewhere in the world they have never been;
- Pursue sometimes vague, conflicting, or incomprehensible objectives;
- Fight an adversary they have never fought;
- Use weapons and equipment they might never have used in combat, in ways that were never intended;
- Operate on the basis of incomplete, untimely, and perhaps incorrect information;
- Execute their orders regardless of weather or visibility;
- Continue to perform in the horrific presence of death or wounding of friendly fighters as well as adversaries;
- Conduct combat operations under the unblinking eye of the television camera and the constant scrutiny of the press;
- Tolerate long separations from family and loved ones;
- Devise ways to deal with the chronic, divisive stresses on internal cohesion and bonding, and on families left behind, caused by the imposition of sex into the shipboard culture;
- Endure lukewarm public support, and at times open hostility, from the home front;
- Absorb casualties from adversaries who fight in unconventional, unanticipated, or illegitimate ways; who have little or no regard for their own personal safety or that of noncombatants; who might be under the influence of performance-enhancing drugs; or who might employ nuclear, chemical, biological, or radiological weapons;

- Achieve their objectives (i.e., win) quickly;
- Inflict minimal numbers of casualties on the adversary;
- Cause minimal levels of destruction to property and the environment and minimal harm to noncombatants, provide assistance to injured combatants and noncombatants alike, and be prepared to restore or reconstruct that which has been damaged;
- Trigger no (or only benign) unintended consequences;

all the while hobbled by guidelines and rules (doctrine, international law, principles of war, rules of engagement) formulated in and for a radically different context. The enormity and the difficulty of such undertakings is, to understate the case, extreme.

Given this list of demanding challenges in a very dynamic context, should it be surprising that the military has great difficulty composing doctrine for combat, peacekeeping, or stabilization operations? Can these tests, with their potential crucial ramifications, be met with less than a coherent, unified, focused, undistracted combat team? Should it be unexpected that the military is often accused of attempting to fight the last war? Should it not be anticipated and accepted that mistakes will be made?

Given this extraordinary list of mostly self-imposed severe challenges in a life-or-death atmosphere, is it not understandable that the military should resist the introduction and perpetuation of sexual dynamism—either by women or homosexuals? Recall that serving in the U.S. military is not a right. The military has long discriminated against those who are too tall, short, fat, thin, young, or old. They also discriminate on the basis of education, criminal record, mental and physical health, physical fitness, and aptitude. Such discrimination does not violate anyone's rights. Should pressure be brought to bear on the military to permit a group of overweight men to serve because they wanted to serve? Yet, there are those who say that women should be allowed to serve in combat units and homosexuals permitted to serve openly merely because they want to. If such integration were to take place, it must be viewed as an experiment, no one can project what the outcome will ultimately be; but one can anticipate with confidence that the problems will be chronic, impossible to completely eliminate, very difficult to adjudicate, and, in the final analysis deleterious to combat readiness.

In short, the military tends, with reason, to believe that there is a positive correlation between combat readiness and victory. Accordingly, anything that will increase the risks of failure in extreme emergencies or combat can be expected to be strongly resisted. And, to restate the obvious, women and homosexuals serving on U.S. Navy ships constitutes a political, not a strategic decision. Strategically, it is wrong; politically, it bends to what a small minority of the populace desire. This is in no way an equivalent trade-off.

Strategy is not solitaire—strategy requires thinking adversaries in order to develop ways to thwart their intentions. But, to the extent that one side is militarily superior to its adversaries in a particular realm, strategy is devalued. It is when adversaries are conjoined in a specific context, either symmetrically or asymmetrically—in which military superiority and technology can tip the balance only with great difficulty—that strategy really matters.

Finding strategies to deal with barbaric asymmetrical adversaries (ones who either won't or can't fight with similar weapons while conforming to the basic rules for the initiation and conduct of armed conflict) will require either non-traditional weapons or finding ways to modify or alter the rules so that they alone do not tilt the battlespace in favor of the bad guys and reduce the ability to accomplish one's objectives: to win. Those who champion the rule of law, arms control, and social experimentation over military pursuit of victory in a just cause have elevated the instruments above the objectives of international politics. Thomas Jefferson had it just right when he wrote that "a strict observance of the written laws is doubtless one of the high duties of a good citizen, but it is not the highest. The laws of necessity, of self-preservation, of saving our country when in danger, are of higher obligation. To lose our country by a scrupulous adherence to written law, would be to lose the law itself, with life, liberty, property and all those who are enjoying them with us; thus absurdly sacrificing the end to the means."[12]

Strategy is fundamental. Without doubt, preparation and professionalism prevent poor performance. Indeed, "Most of the time on board a naval vessel was spent in eternal preparation for a situation—be it collision at sea, man overboard, a fire in the powder room, or war—that might never take place. Drills, routine upkeep, and watches filled the naval day. Even in wartime, the only activity for months might be the steady motion of the ship."[13]

Preparation and professionalism must eternally be linked closely to context. Even a cursory review of the foregoing list of demands on military forces demonstrates the primacy of context. All are intimately related to context: place, adversary, weapon selection by both sides, orders, weather, casualties, information, objectives, news coverage, family separation, public support, laws and rules, collateral damage, and unintended consequences.

Over the course of many years the United States has come to rely on its naval forces to assist not only in defense of the homeland, but also to oppose threats to the international order: cross-border aggression, massive violations of human rights, imbalances of power, proliferation of weapons of mass destruction, drug trafficking, and transnational terrorism. The adoption and refinement of the strategic culture of the U.S. Navy over the centuries has prepared it well to deal with threats and constraints alike. Forged as it was at sea; tempered by wind, storm, and battle; conditioned by interminable hours searching, searching, searching for the adversary; honed by the pressing, continuous needs of systems (whether of sails and rigging, holystones, ground tackle, forced-draft blowers, primary coolant loops, evaporators, hydraulic pumps, or electronic circuits); fortified by extended voyages in remote areas of the boundless seas; and grounded both in history and experience with conceptual, systematic, non-linear, adaptive thought processes.

From the foregoing, it now appears "incontestable that the Navy long has produced our military's best strategic thinkers—captains and admirals able to transcend parochial interests to see the global security environment as a whole."[14] And this would seem to close the circle on how this book began. Of course, during the course of its writing, Admiral Fallon resigned his post as Central Command commander, reportedly because his strategic vision differed from that of the National Command Authorities. It is possible that Admiral Fallon forgot a bit of his history, for "in the course of America's fight against the Barbary terror, the nation tried appeasement, diplomacy, the threat of force, and war. Although the United States settled for peace in 1805, it seemed clear that war, or at least the judicious and credible threat of force, was far more eloquent a method of persuasion with Barbary than money or goodwill."[15] On the other hand, perhaps he was thinking about the longer term, recognizing that "broadly speaking, sea power works slowly and subtly, whereas generals, politicians, and the people at large are impatient for direct, immediately apparent results, as with armies on the march."[16] And, conceivably, the entire business was overblown—inflated idle gossip bandied about by a third party.[17]

In a book published over sixty years ago, Bernard Brodie—one of the great post–World War II American strategists—wrote, "Foreign policy tends to run along channels determined by tradition, and Powers which have been great are considered great until they are proved otherwise."[18] What pertains to foreign policy pertains to sea power in spades.

Time is linear: time "marches on." It is not possible to rewind time, set a different course, and see if the journey and destination are better than the one chosen earlier. People must live with the consequences of their actions, and learn from them. "Those who cannot remember the past are condemned to repeat it," or so the bumper sticker reads.

Yet, as with all bumper stickers, one must ask, "What is the rest of the story?" And, more often than not, the "rest of the story" contains the real message. Here is what George Santayana wrote:

> Progress, far from consisting in change, depends on retentiveness. When change is absolute there remains no being to improve and no direction is set for possible improvement: and when experience is not retained, as among savages, infancy is perpetual. Those who cannot remember the past are condemned to repeat it. . . . In a moving world readaptation is the price of longevity.[19]

In this passage Santayana has described the power of Navy Strategic Culture. He writes that humankind makes progress because it learns from its experience and retains its knowledge, which it then applies to current and future situations. Change, he avers, depends on remembering—otherwise one invariably acts as a savage or an infant. But then, once retentiveness has run its course and forgetfulness begins to prevail, people remake the same mistakes because they fail to adapt.

Navy Strategic Culture draws on a rich trove of history and tradition, steeped in the never-ending struggle with Mother Nature and conditioned by a harsh, unstructured environment in which the politics of humans play only a miniscule part. It leavens that collective memory with cumulative experience. In the application of that collected and tempered knowledge, it is agile and adaptive, permitting it to live and to move forward in new, solidly anchored ways. The future of the United States of America depends importantly on the continued longevity of its Navy's Strategic Culture.[5]

Appendix: Treasure Chest of Quotations

Chapter 1: Introduction

"Historians generally have been unfamiliar with the conditions of the sea, having as to it neither special interest nor special knowledge; and the profound determining influence of maritime strength upon great issues has consequently been overlooked." Alfred Thayer Mahan, *The Influence of Sea Power upon History, 1660–1783* (London: Methuen and Company, 1965; first pub. 1890), v.

"For whether he was rich or poor, fisherman or pirate, every West Country man shared a distinct maritime culture, a world strikingly different from that which arises from the routines of the land. It had its own values, its own outlook on life, even its own language—as sailors still do today. For landlubbers thrown into this peculiar world, the effect could be disconcerting. Ancient philosophers believed seafarers formed their own category of beings distinct from either the living or the dead." Arthur Herman, *To Rule the Waves: How the British Navy Shaped the Modern World* (New York, NY: Harper Perennial, 2004), 28.

" 'If you get five Navy officers in a room to solve a given problem, the answer is going to be different than if you had five Army officers and five Navy officers.' That can have a significant impact when it comes to considering an extended stay in Iraq and Afghanistan or what post-Iraq defense strategy ought to resemble."

Robert Work, a retired Marine colonel, quoted in Spencer Ackerman, "Anchors Aweigh! Behind the Supremacy of the Navy," *Talking Points Memo*, June 8, 2007.

Without a Navy, "a nation, despicable by its weakness, forfeits even the privilege of being neutral." Alexander Hamilton, "The Utility of the Union in Respect to Commercial Relations and a Navy," Federalist 11, in James Madison, Alexander Hamilton, and John Jay, *The Federalist Papers* (New York, NY: Penguin Classics, 1987), 131.

"The dependence of the American people upon the sea—to utilize it for political and economic power and even for cultural growth—has been so dramatically demonstrated throughout our history as to make the image of the republic incomprehensible without it. Without American activity upon the waters, alone and with allies, how different the course of American and world history would have been—from the crossing of the Pilgrim fathers on the Mayflower to the recovery of the Apollo astronauts. Clark G. Reynolds, "The Sea in the Making of America," *To Use the Seas: Readings in Seapower and Maritime Affairs* (Annapolis: Naval Institute Press, c. 1977), 31.

"The airman conquers his environment; the sailor survives it . . . even the release of surrender is beyond the reach of the sailor; he fights and dies with his ship. . . . These forces imbue the sailor with a unique combination of qualities: self-reliance, a special respect and regard for the person who is in charge of his vessel and absolute accountability. . . . The limited confines of a warship—even a very large one—force close associations among its inhabitants." James A. Winnefeld, "Why Sailors Are Different," U.S. Naval Institute *Proceedings* (May 1995): 66.

"Where the sailor and the airman are almost forced, by the nature of the sea and the air, to think in terms of a total world or, at the least, to look outside the physical limits of their immediate concerns, the soldier is almost literally hemmed in by his terrain." J. C. Wylie, *Military Strategy: A General Theory of Power Control* (Annapolis, MD: Naval Institute Press, c. 1967), 42.

"If the United States shall determine to augment their navy, so as to rival those of Europe, the public debt will become permanent; direct taxes will be perpetual; the paupers of the country will be increased; the nation will be bankrupt; and, I fear, the tragedy will end in a revolution." Representative Adam Seybert of Pennsylvania, 1811, quoted in Marion Mills Miller, ed., *Great Debates in American History*,

(New York, NY: Current Literature Publishing Company, 1913). http://www.shsu.edu/~his_ncp/Nav1812.html. (September 15, 2007).

"The sailors standing ready on the quay looked at the soldiers marching up curiously, with something of pity and something of contempt mingled with their curiosity. The rigid drill, the heavy clothing, the iron discipline, the dull routine of the soldier were in sharp contrast with the far more flexible conditions in which the sailor lived." C. S. Forester, *Mr. Midshipman Hornblower* (New York, NY: Pinnacle Books, 1948), 123.

"In placing greater emphasis on art than on science when it came to the question of command in war, the Mahans [father and son] resembled Clausewitz rather than Jomini. This was especially true with regard to attitude toward comprehensive theory." Jon Tetsuro Sumida, *Inventing Grand Strategy and Teaching Command: The Classic Works of Alfred Thayer Mahan Reconsidered* (Washington, D.C.: The Woodrow Wilson Center Press, and Baltimore: The Johns Hopkins University Press, 1997), 112.

"The maritime strategies are the one field in which the United States has an inherent advantage over any enemy. The sailor hopes the nation, if it is ever forced to war, will take advantage of that, use it, and exploit it for all it is worth." Wylie, *Military Strategy,* 162.

"Reflecting on the Gulf of Tonkin incident, former White House aide Chester Cooper observed: 'There is something very magical about an attack on an American ship on the high seas. An attack on a military base or an Army convoy doesn't stir up that kind of emotion. An attack on an American ship on the high seas is bound to set off skyrockets and the 'Star-Spangled Banner' and 'Hail to the Chief' and everything else.' " Sean M. Lynn-Jones, "A Quiet Success for Arms Control: Preventing Incidents at Sea," *International Security* (Spring 1985): 164.

The Lord gave us two ends to use
One to think with and one to sit with
The outcome depends on which way we choose
Heads we win—tails we lose.
Chester Nimitz, quoted in Charles Lee Lewis, *Famous American Naval Officers* (Manchester, NH: Ayer Publishing, 1971), 426.

"Imagination is more important than knowledge. For knowledge is limited, whereas imagination embraces the entire world." Albert Einstein. http://www.einsteinyear.org/facts/physicsFacts/ (January 9, 2008).

"The art of the sailor is to leave nothing to chance." Annie Van De Wiele. http://www.svsouthernstar.com/nauticalquotes.php (January 9, 2008).

"Chance favors the prepared mind." Louis Pasteur. http://www.quotedb.com/quotes/2195 (January 9, 2008).

"The wonder is always new that any sane man can be a sailor." R. W. Emerson in Robert Debs Heinl Jr., *Dictionary of Military and Naval Quotations* (Annapolis, MD: Naval Institute Press, 1966), 285.

"Public debate and scholarly literature are not studded with references to 'strategic' sea power, but by its very nature sea power is strategic in its working." Colin S. Gray, *The Navy in the Post-Cold War World: The Uses and Value of Strategic Sea Power* (University Park, PA: The Pennsylvania State University Press, 1994), 9.

"We don't fight for land; we don't fight for money; and we won't fight to simply impose our will. But we will take to arms for the principle of freedom." Remarks of the Secretary of the Navy The Honorable Gordon R. England at the Navy Birthday Ball, Naval District Washington, Washington, D.C., October 12, 2002. http://www.navy.mil/navydata/people/secnav/england/speeches/eng-ndwbd.txt (August 15, 2008).

"Sea power is not about the military effect of fighting ships; rather, it is about the use of maritime lines of communication for the effective interconnection, organization, and purposeful application of the warmaking potential of many lands." Colin S. Gray, *War, Peace, and Victory* (New York, NY: Simon and Schuster, 1990), 75.

Chapter 2: Strategic Culture

"Gentlemen, you can no more make a sailor out of a land-lubber by dressing him up in sea-toggery and putting a commission into his pocket, than you could make a showmaker of him by filling him up with sherry cobblers." David Farragut, quoted in Peter Karsten, *The Naval Aristocracy: The Golden Age of Annapolis and the Emergence of Modern American Navalism* (New York, NY: The Free Press, 1972), 33.

"Consider these words of Napoleon Bonaparte: 'Tactics, evolutions, artillery, and engineer sciences can be learned from manuals like geometry; but the knowledge of the higher conduct of war can only be acquired by studying the history of wars and the battles of great generals and by one's own experience.' " Colin S. Gray, "Why Strategy is Difficult," *Joint Force Quarterly* (Summer 1999): 10.

"Wars are conceived, plotted, and waged by socially conditioned human agents." R. Craig Nation, "Regional Studies in a Global Age," in J. Boone Bartholomees Jr., ed., *U.S. Army War College Guide to National Security Policy and Strategy* (July 2004), 59.

"Military culture is shaped by national cultures as well as factors such as geography and historical experience that build a national military 'style.' . . . Changes in leadership, professional military education, doctrinal preference, and technology all result in the evolution, for better or worse, of the culture of military institutions." Williamson Murray, "Military Culture Does Matter," *FPRI Wire* 7, no. 2 (January 1999).

"A man can become a great captain only with a passion for study and a long experience. There is not enough in what one has seen oneself; for what life of a man is fruitful enough in events to give a universal experience; and who is the man that can have the opportunity of first practicing the difficult art of the general before having filled that important office? It is, then, by increasing one's own knowledge with the information of others, by weighing the conclusions of one's predecessors, and by taking as a term of comparison the military exploits, and the events with great results, which the history of war gives us, that one can become skillful therein." Archduke Charles of Austria, quoted in John B. Hattendorf, ed., *Mahan on Naval Strategy: Selections from the Writings of Rear Admiral Alfred Thayer Mahan.* (Annapolis, MD: Naval Institute Press, 1991), 276.

"The key to performance under pressure is real knowledge gained from extensive study and intensive training. Such skills require time to develop, are fragile, and decay quickly when not used. . . . Long and repeated tours are necessary in any part of military trade that requires detailed knowledge and developed leadership." Rear Admiral W. J. Holland, Jr., USN (Retired), "Jointness Has Its Limits," *Proceedings* (May 1993), 42.

"Historians note that through most of American history there has been disdain for military service among the general population. . . . The greatest fear, harbored right at the outset by the new nation, was that the military might break loose from the bonds of civilian control." Robert Bateman, "The Army and Academic Culture," *Academic Questions* (Winter 2007–2008): 64.

"History is simply recorded memory. People without memory are mentally sick. So too are nations or societies or institutions that reject or deny the relevance of their collective past." Philip A. Crowl, "The Strategist's Short Catechism: Six Questions without Answers," in Harry R. Borowski, ed., *The Harmon Memorial Lectures in Military History, 1959–1987* (Washington, D.C.: Office of Air Force History, 1988), 377.

" 'Now, it must strike anyone who thinks about it as extraordinary,' Luce also said, 'that we, members of a profession of arms, should never have undertaken the study of our real business—war.' " Quoted in Russell F. Weigley, *The American Way of War: A History of United States Military Strategy and Policy* (Bloomington, IN: Indiana University Press, 1973), 172.

"The sea demands definite qualities in the seafarer—certain attitudes of mind and character. Humility, prudence, and a recognition that there is no end to learning and to the acquisition of experience." Alasdair Garrett, source unknown.

"War and its cognate branches constitute the college curriculum. It is only by a close study of the science and art of war that we can be prepared for war, and thus go very far toward securing peace." Stephen B. Luce, quoted in John B. Hattendorf, B. Mitchell Simpson III, and John R. Wadleigh, *Sailors and Scholars: The Centennial History of the U.S. Naval War College* (Newport, RI: Naval War College Press, 1984), 22.

"Regular [definitional] and irregular [conceptual] forms coexist but require different computational mechanisms: symbol combinations for regular forms, associative memory for irregular forms. The same may be true for classical and family resemblance categories." Steven Pinker, *Words and Rules: The Ingredients of Language* (New York, NY: Basic Books, 1999), 278.

"Men who deliberately postpone the formation of opinion until the day of action, who expect from a moment of inspiration the results commonly obtained only from study and reflection, who hope for victory in ignorance of the rules that have generally given victory, are guilty of yet greater folly, for they disregard all the past experience of our race. . . . 'Upon the field of battle,' says the great Napoleon, 'the happiest inspiration is most often only a recollection.' " Hattendorf, *Mahan on Naval Strategy*, xxix.

"Study of the workings of the minds of the proved strategists stimulates, if it does not actually procreate, ideas. It opens up one's mental vision, it widens one's strategic horizon. . . . We see, in history, how the Masters of War have tackled their problems. Would it not be something approaching impudence to pretend that we can learn nothing from them, that we are self-sufficient in ourselves?" Herbert Richmond, c. 1928, source unknown.

"Metaphor is pervasive in everyday life, not just in language but in thought and action. . . . The concepts that govern our thoughts are not just matters of the intellect. They also govern our everyday functioning, down to the most mundane details. Our concepts structure what we perceive, how we get around in the world, and how we relate to other people . . . the way we think, what we experience, and what we do every day is very much a matter of metaphor." George Lakoff and Mark Johnson, quoted in Peter Morville, *Ambient Findability* (Sebastopol, CA: O'Reilly, 2005), 33.

"History provides a powerful antidote to contemporary arrogance." Paul Johnson, quoted in Thomas Sowell, *Basic Economics: A Common Sense Guide to the Economy*, 3d. ed. (New York, NY: Basic Books, 2007), 547.

"Naval Officer: An upright, God-fearing man, not dainty about his food or drink, robust and alert, with good sea-legs, and in strong voice to give commands to all hands; pleasant and affable in conversation, but imperious in his commands, liberal and courteous to defeated enemies, knowing everything that concerns the handling of the ship." Samuel de Champlain, *Treatise on Seamanship*, quoted in Heinl, *Dictionary*, 206.

"Inherently armies are deeply ritualistic organisations, and some of their ritual is devoted to the marking of important events in the individual's service—his oath of enlistment, passing-out from recruit training, and so on." Richard Holmes, *Acts of War: The Behavior of Men in Battle* (New York, NY: The Free Press, 1985), 28.

"People of common sense have a good instinct for which bit of wisdom fits which context. But you might tie them up in knots if you ask them for a theory that irons out all the edges and wrinkles in what they say. They don't think theoretically. They try to take the facts in context, case by case, and reach back in memory for the appropriate wisdom. They do not look for a theory, but for a ray of practical wisdom helpful just now." Michael Novak, *On Two Wings: Humble Faith and Common Sense at the American Founding,* expanded ed. (San Francisco, CA: Encounter Books, 2002), 108.

"The French word doctrinaire, fully adopted into English gives warning of the danger that attends doctrine; a danger to which all useful conceptions are liable. The danger is that of exaggerating the letter above the spirit, of becoming mechanical instead of discriminating. This danger inheres especially in—indeed, is inseparable from—the attempt to multiply definition and to exaggerate precision; the attempt to make a subordinate a machine working on fixed lines, instead of an intelligent agent, imbued with principles of action, understanding the general character not only of his own movement, but of the whole operation of which he forms part; capable, therefore, of modifying action correctly to suit circumstances." Alfred Thayer Mahan, *Mahan on Naval Strategy: Selections from the Writings of Rear Admiral Alfred Thayer Mahan* (Annapolis, MD: Naval Institute Press, 1991), 349.

"A modern navy is a totally untried weapon of warfare. It is the resultant of a host of more or less conflicting theories of attack and defence." Winston S. Churchill, *The Second World War*, vol. 4 (London: Cassell, 1954), 554.

"The Spartan team had a corporate permanence, and this is one of the first of the great gifts that Sparta left the world: the idea that institutions did not depend on the particular individuals that happened to make them up at any particular time, but that their composition could change—and change completely—without the essential nature of the team changing in the least. . . . Spartans deliberately worked to make themselves individually fungible—interchangeable—and this is why they died not for personal glory or honor but from a sense of duty." Harris, *Civilization and Its Enemies,* 91–92.

"In military usage, the term 'cohesion' describes the bonding of a unit's fighters that converts it from a group into a force. A cohesive force functions as a unified, interconnected system. . . . Cohesion offers structural, psychological, and purposeful integrity. Like a formation of ships, a cohesive force looks and behaves as if it were a single organism. . . . There are two determinants of cohesion . . . one is structural, the other behavioral. In military affairs, structure is defined by organization, chain of command, lateral coordination, and the choice of equipment at each echelon. . . . Whereas structural cohesion comes in large part from doctrinal prescriptions of beliefs and arrangements, behavioral cohesion derives from doctrinal choices of action. . . . Bonding all participants with faith in their united skills makes maneuver warfare—or any other soundly conceived style—effective." Wayne P. Hughes Jr., "The Power in Doctrine," *Naval War College Review* vol. 48, no. 3 (Summer 1995): 12–14.

"The military depends upon men acting as a team at the very moment when every man is under great temptation to seek his own safety. The personal bonds that men form with each other as leaders, as followers, as comrades-in-arms, often enable ordinary men to perform acts of extreme self-sacrifice when ideals such as duty, country, or cause no longer compel." Brian Mitchell, *Women in the Military: Flirting with Disaster* (Washington, D.C.: Regnery, 1998), 174.

"Individuals do not fight wars; groups do . . . the effectiveness of a combat force is not simply the sum of the attributes of individuals but rather a product of how the group works together." Kingsley Browne, *Co-Ed Combat: The New Evidence That Women Shouldn't Fight the Nation's Wars* (New York, NY: Sentinel, 2007), 12

"Much of men's motivation to fight comes from their appreciation of the link between warfare and masculinity, and disruption of that link is likely to diminish their motivation. . . . Men fight for many reasons, but probably the most powerful one is the bonding—male bonding—with their comrades." Browne, *Co-Ed Combat,* 7.

"The absence of even a few men from a shipboard fire party or a machinery repair group can have a serious effect on the outcome of an emergency situation at sea. . . . At sea, sailors share an implicit social contract of social interdependence. Few state the obvious, but for the sleeping crew members below-decks have forfeited control of their very survival because they trust the officer of the deck and his watch." John B. Bonds, "Punishment, Discipline, and the Naval Profession," U.S. Naval Institute *Proceedings* (December 1978): 47.

"Nelson understood that the fog of war was not a cloud of uncertainty that could be whisked away if only the appropriate communicative means were at hand, but an element inherent to warfare as life itself: an element as resistant to the dissipatory effect of commanders as real fog was to the blows of a sword." Michael A. Palmer, *Command at Sea: Naval Command and Control since the Sixteenth Century* (Cambridge, MA.: Harvard University Press, 2005), 321–322.

"Clausewitz, like the Mahans, believed that command had to be exercised in the face of uncertainty and thus required moral as well as intellectual qualities." Sumida, *Inventing Grand Strategy*, 112.

"The friction of war cannot be duplicated or simulated adequately in games, studies and exercises. 'We have identified danger, physical exertion, intelligence, and friction as the elements that coalesce to form the atmosphere of war, and turn it into a medium that impedes activity. In their restrictive effects they can be grouped into a single concept of general friction. Is there any lubricant that will reduce this abrasion? Only one, and a commander and his army will not always have it readily available: combat experience.' " Carl von Clausewitz, *On War*, bk. 1, Michael Howard and Peter Paret, eds. (Princeton, NJ: Princeton University Press), 122.

" [J. Glenn] Gray cited Jesus' comment that 'Greater love has no man than this, that a man lay down his life for his friends.' Gray then followed Jesus' thought with a question: 'What meaning has friendship for warriors?' He answered it with a thoughtful examination, essentially contending that there is a moment when warriors become conscious of the fact that, 'I am part of you, and you are part of me.' Warriors, therefore, see themselves in another and recognize that friendship is about a level of human interrelatedness and connectedness not fully known before." Steven L. Smith, "Crossing Swords: Let Us Pray," U.S. Naval Institute *Proceedings* (January 2007): 21–23.

"Fear is the common bond between fighting men. The overwhelming majority of soldiers experience fear during or before battle: what vary are its physical manifestations, its nature and intensity, the threat which induces it, and the manner in which it is managed." Holmes, *Acts of War*, 204.

In the words of David Porter, the captain of the ship has, ". . . no society, no smiles, no courtesies for or from anyone. Wrapped up in his notions of his own dignity, and the means of preserving it, he shuts himself up from all around him. He stands alone, without the friendship or sympathy of one on board; a solitary being in the midst of the ocean." Quoted in Christopher McKee, *Edward Preble: A Naval Biography, 1761–1807* (Annapolis, MD: Naval Institute Press, 1972), 318–319.

Admiral Halsey to Admiral Burke before the Battle of Cape Saint George: "If enemy contacted, you know what to do." Quoted in The Honorable Jerry MacArthur Hultin, "The Business Behind War Fighting," United States Naval Institute *Proceedings* (September 1999): 59.

"[I]n acting to provide security for American business interests abroad, the Navy came into hostile contact with armed forces of foreign states on at least 44 different occasions between 1865 and 1927. . . not counting the instances when force was merely threatened." Karsten. *Naval Aristocracy,* 168.

On tradition: "In the politics of power, military prestige is the medium of account, and nothing gives a nation greater prestige than past military victories. The political, economic, and most of all technological conditions under which those victories were won may have changed, but this counts for little except among the most discerning few. Foreign policy tends to run along channels determined by tradition, and Powers which have been great are considered great until they are proved otherwise." Bernard Brodie, *Sea Power in the Machine Age* (New York, NY: Greenwood Press, 1943), 431.

Many things we find around us could not be deduced by any body of laws, because they are shaped by myriad events of history no longer visible to us." Pinker, *Words and Rules,* 282.

"The cat-o'nine tails was a most hated instrument of punishment, the cathead, catfall, catblock and catdavit were various pieces of gear associated with the laborious task of raising anchor, the catlap referred to a watered-down, or weak drink, a catnap was an inadequate sleep, and a cat's paw was an insufficient breath of wind." James Clary, *Superstitions of the Sea* (St. Clair, MI: Maritime History in Art, 1994), 8.

How little do the landsmen know
Of what we sailors feel,
When waves do mount and winds do blow!
But we have hearts of steel.
"The Sailor's Resolution" in Heinl, *Dictionary,* 284.

"It is not the critic that counts; not the man who points out how the strong man stumbles or the doer of deeds could have them better. The credit belongs to the man who is actually in the Arena, whose face is marred by dust and sweat and blood; who strives valiantly; who errs and comes short again and again, because there is no effort without error and shortcoming; but he who does actually strive to do the deed; who knows the great devotion; who spends himself in a worthy cause, who at the best, knows in the end the triumph of high achievement, and who at the worst, if he fails while daring greatly, knows that his place shall never be with those cold and timid souls, who know neither victory nor defeat." Theodore Roosevelt. http://www.fairtax.net/quotes.htm.

"Of all careers, the navy is the one which offers the most frequent opportunities to junior officers to act on their own." Napoleon, *Political Aphorisms*, 1848, quoted in Heinl, *Dictionary*, 209.

"An admiral can influence personally only the men on the vessel on which he finds himself; smoke prevents signals from being seen and winds change or vary over the space occupied by his line. It is thus of all professions that in which subalterns should use the largest initiative." *Napoleon's Maxims*, CXV. http://www.pvv.ntnu.no/~madsb/home/war/napoleon/maxim02.php (August 15, 2008).

The Old Outfit
written by a World War Two Sailor

Come gather round me lads and I'll tell you a thing or two,
About the way we ran the Navy in nineteen forty two.
When wooden ships and iron men were barely out of sight,
I am going to give you some facts just to set the record right.
We wore the ole bell bottoms, with a flat hat on our head,
And we always hit the sack at night. We never "went to bed."
Our uniforms were worn ashore, and we were mighty proud.
Never thought of wearing civvies, in fact they were not allowed.
Now when a ship puts out to sea. I'll tell you son—it hurts!
When suddenly you notice that half the crews wearing skirts.
And it's hard for me to imagine, a female boatswains mate,
Stopping on the Quarter deck to make sure her stockings are straight.
What happened to the KiYi brush, and the old salt-water bath
Holy stoning decks at night—cause you stirred old Bosn's wrath!

We always had our gedunk stand and lots of pogey bait.
And it always took a hitch or two, just to make a rate.
In your seabag all your skivvies, were neatly stopped and rolled.
And the blankets on your sack had better have a three-inch fold.
Your little ditty bag . . . it is hard to believe just how much it held,
And you wouldn't go ashore with pants that hadn't been spiked and belled.
We had scullery maids and succotash and good old S.O.S.
And when you felt like topping off—you headed for the mess.
Oh we had our belly robbers—but there weren't too many gripes.
For the deck apes were never hungry and there were no starving snipes.
Now you never hear of Davey Jones, Shellbacks Or Polliwogs,
And you never splice the mainbrace to receive your daily grog.
Now you never have to dog a watch or stand the main event.
You even tie your lines today—back in my time they were bent.
We were all two-fisted drinkers and no one thought you sinned,
If you staggered back aboard your ship, three sheets to the wind.
And with just a couple hours of sleep you regained your usual luster.
Bright eyed and bushy tailed—you still made morning muster.
Rocks and shoals have long since gone, and now it's U.C.M.J.
THEN the old man handled everything if you should go astray.
Now they steer the ships with dials, and I wouldn't be surprised,
If some day they sailed the damned things—from the beach computerized.
So when my earthly hitch is over, and the good Lord picks the best,
I'll walk right up to HIM and say, "Sir, I have but one request—
Let me sail the seas of Heaven in a coat of Navy blue.
Like I did so long ago on earth—way back in nineteen-forty two."
Proudly copied from Lt. (jg) Don Ballard, USN (Ret.), April 13, 2002, who loved the Navy and all the men he served with in all of World War Two. Courtesy of the author's son. http://www.goatlocker.org/resources/nav/outfit.htm (August 16, 2008).

"There are three sorts of people; those who are alive, those who are dead, and those who are at sea." Old Capstan Chanty attributed to Anacharsis, sixth century BC.

"The social separation from society is most effectively achieved through the use of institutionally specific language." Alastair Finlan, *The Royal Navy in the Falklands Conflict and the Gulf War: Culture and Strategy* (London: Frank Cass, 2004), 6.

The Laws of the Navy
written by Adm. R. A. Hopwood, RN

Now these are the laws of the Navy,
Unwritten and varied they be;
And he who is wise will observe them,
Going down in his ship to the sea.

As naught may outrun the destroyer,
So it is with the law and its grip,
For the strength of a ship is the Service,
And the strength of the Service the ship.

Take heed what you say of your seniors,
Be your words spoken softly or plain,
Let a bird of the air tell the matter,
And so shall ye hear it again.

If you labour from morn until even,
And meet with reproof for your toil,
'Tis well, that the gun may be humbled
The compressor must check the recoil.

On the strength of one link in the cable,
Dependeth the might of the chain.
Who knows when thou may'st be tested?
So live that thou bearest the strain!

When a ship that is tired returneth,
With the signs of the seas showing plain;
Men place her in dock for a season,
And her speed she reneweth again.

So shall ye, if perchance ye grow weary,
In the uttermost parts of the sea,
Pray for leave, for the good of the Service,
As much and as oft as may be.

Count not upon certain promotion
But rather to gain it aspire;
Though the sightline may end on the target
There cometh perchance the miss-fire.

Can'st follow the track of the dolphin?
Or tell where the sea swallows roam?
Where Leviathan taketh his pastime?
What ocean he calleth his own?

So it is with the words of the rulers,
And the orders these words shall convey;
Every law is naught beside this one:
Thou shalt not criticise, but Obey.

Say the wise: How may I know their purpose?
Then acts without wherefore or why.
Stays the fool but one moment to question,
And the chance of his life passes by.

If ye win through an African jungle,
Unmentioned at home in the press,
Heed it not. No man seeth the piston,
But it driveth the ship none the less.

Do they growl? it is well. Be thou silent,
If the work goeth forward amain.
Lo! the gun throws the shot to a hair's breadth
And shouteth, yet none shall complain.

Do they growl, and the work be retarded?
It is ill, be whatever their rank.
The half-loaded gun also shouteth,
But can she pierce target with blank?

Doth the paintwork make war with the funnels
And the deck to the cannons complain?
Nay, they know that some soap and fresh water
Unites them as brothers again.

So ye, being heads of departments,
Do you growl with a smile on your lip,
Lest ye strive and in anger be parted,
And lessen the might of your ship.

Dost deem that thy vessel needs gilding,
And the dockyard forbears to supply?
Put thy hand in thy pocket and gild her—
There are those who have risen thereby.

Dost think in a moment of anger
'Tis well with thy seniors to fight?
They prosper, who burn in the morning,
The letters they wrote overnight.

For many are shelved and forgotten,
With nothing to thank for their fate,
But that on a half sheet of foolscap
A fool "Had the honour to state."

Should the fairway be crowded with shipping
Beating homeward the harbour to win,
It is meet that lest any should suffer,
The steamers pass cautiously in.

So thou, when thou nearest promotion,
And the peak that is gilded is nigh,
Give heed to words and thine actions,
Lest others be wearied thereby.

It is ill for the winners to worry,
Take thy fate as it comes, with a smile,
And when thou art safe in the harbour
They may envy, but will not revile.

Uncharted the rocks that surround thee,
Take heed that the channels thou learn,
Lest thy name serve to buoy for another
That shoal the "Court-Martial Return."

Though a Harveyised belt may protect her
The ship bears the scar on her side;
'Tis well if the Court should acquit thee—
But 'twere best had'st thou never been tried.

MORAL

As the wave washes clear at the hawse pipe,
Washes aft, and is lost in the wake;
So shalt thou drop astern all unheeded
Such time as these laws ye forsake.

Take heed in your manner of speaking
That the language ye use may be sound,
In the list of the words of your choosing
"Impossible" may not be found.

Now these are the Laws of the Navy,
And many and mighty are they.
But the hull and the deck and the keel
And the truck of the law is—OBEY.
Naval Historical Center. http://www.history.navy.mil/photos/arttopic/titles/law-navy.htm (August 16, 2008).

"'Culture' . . . implies a way of life that has been learned from the social and physical environment, as opposed to one that has been inherited." Donald B. Calne, *Within Reason: Rationality and Human Behavior* (New York, NY: Vintage Books, 1999), 67.

"The human ability to learn from experience and nature, so slighted in current humanistic theory, is not merely an object of cultural transmission, let alone of social control, but an evolutionary triumph of the species, indeed, a triumph on which our future ultimately depends. . . . If the role of both acquired knowledge and the transmission and emendation of the means of acquiring knowledge is a 'Western' concern, then it is a Western concern upon which human fate depends. . . . The demand for perfection is antinomian, illogical, and empirically absurd. The triumph of the West is flawed but real." Kors, "The West," 4.

"Our values are not matters of whim and happenstance. History has given them to us. They are anchored in our national experience, in our great national documents, in our national heroes, in our folkways, traditions, and standards. People with a different history will have differing values. But we believe that our own are better for us. They work for us; and, for that reason, we live and die by them." Arthur M. Schlesinger Jr., *The Disuniting of America* (New York, NY: W. W. Norton & Company, 1992), 137.

"From the earliest days of its founding, the nation has been guided by a philosophy that social historians call the American Creed. The creed's paramount values are self-reliance, stoicism, courage in the face of adversity, and the valorization of excellence." Christina Hoff Sommers and Sally Satel, *One Nation Under Therapy* (New York, NY: St. Martin's Press, 2005), 218.

Chapter 3: The Maritime Context

"Which brings us to graffiti theory, my corollary to both memory-prediction and broken windows, which suggests that all information that flows through our senses continuously and unconsciously shapes our memories, beliefs, predictions, decisions, and behaviors. We are born with instinct, but in matters of intuition, we are lifetime learners. Information is data that makes a difference, literally." Morville, *Ambient Findability*, 169.

"How inappropriate to call this planet Earth when it is clearly Ocean." Arthur C. Clark. http://thinkexist.com/quotation/how-inappropriate-to-call-this-planet-earth-when/407084.html (August 15, 2008).

"Those who have never seen themselves surrounded on all sides by the sea can never possess an idea of the world and their relation to it." Johann Wolfgang von Goethe, *Many Colored Threads from the Writings of Goethe* (Boston, MA: Lothrop Company, 1885), 89.

"[The sea] . . . an immensity that receives no impress, preserves no memories, and keeps no reckoning of lives." Joseph Conrad, quoted in James Tazelaar, ed., *The Articulate Sailor* (Tuckahoe, NY: John de Graff Inc., 1973), 64.

"On 17th December 1944, a fleet of over 80 American warships (destroyers, cruisers and aircraft carriers) in the Pacific received warning of an incoming typhoon. The ships' companies made the preparations they could but after an unequal struggle, three destroyers had capsized and sunk, several other ships were severely damaged, and nearly 800 lives had been lost." Robert D. G. Webb and Tabbeus M. Lamoureux, *Human Reliability and Ship Stability* (Guelph, ON: Humansystems Incorporated, July 4, 2004), 2.

Roll on, thou deep and dark blue Ocean—roll!
Ten thousand fleets sweep over thee in vain;
Man marks the earth with ruin—his control
Stops with the shore.
Lord Byron, in *The Oxford Dictionary of Quotations,* 3d ed. (Oxford: Oxford University Press, 1979), 121.

"As these sailors ventured farther out to sea, they were not only vulnerable to the real dangers of freak weather phenomena in their fragile boats but they also sailed with the imaginary fears of sea monster, fiends, devils, dragons, gods of monstrous shapes, fire-breathing bulls, terrific giants, enchanting sirens, dwarfed pygmies, man-eaters, seas of darkness, and rogue waves. The unknown bred fear." Clary, *Superstitions,* 2.

"One of the most renowned omens is that of Saint Elmo's fire, or corposant, the fiery light that often appears at the tips of mastheads, yardarms, stays, etc. The glowing discharge, often accompanied by a crackling sound, takes place when the air in a storm is charged with electricity." Clary, *Superstitions*, 220.

"We must never forget, in our enthusiasm for the new and the spectacular means of transport, that nearly 70 percent of the earth's surface is covered by water. We must never forget as Americans, that this continent is an island situated between two great oceans whose surface or air-space we cannot dare to leave unguarded." John L. Sullivan, undersecretary of the Navy, in speech in New York, November 15, 1946. Courtesy Naval Historical Center, Washington, D.C.
"The most common characteristic of sea water . . . is its saltiness. Striking a general average . . . every 1,000 pounds of sea water carries in solution 35 pounds of solid matter, of which more than 27 pounds is sodium chloride, or common salt. . . . Actually, the sea is the ultimate source of all salt, for the origin of salt deposits can invariably be traced back to the drying up of various seas." Coote, *Norton Book of The Sea*, 4–5.

"The absence of a single familiar reference point can be utterly unnerving even to the coolest and clearest minds." http://www.scribd.com/doc/3197363/Arthur-C-Clarke-Childhoods-End (September 11, 2007).

"Those who live by the sea can hardly form a single thought of which the sea would not be part." Hermann Broch, quoted at http://www.quotecosmos.com/quotes/35288/view (June 20, 2008).

"Because the sea lives (while the land lies inert) we cannot think of it as dumb: nor is it. But it speaks in a veiled fashion as do the oracles of the Gods, whereof it is one." Hilaire Belloc, quoted in Tazelaar, *The Articulate Sailor*, 67.

"I took over at 0800. Conditions were absolutely shocking: the sort of thing one reads about but does not believe. A wind that has reached a state of senseless fury: a wind that soon numbs one into a dull state of hopelessness: a kind of absolute demon force that piles the sea into unstable toppling heaps. And with the wind came at frequent intervals, the most blinding rain I have ever seen. It was impossible to face the wind and open one's eyes . . . sea and air had become inextricably mingled." Humphrey Barton, quoted in Tazelaar, *The Articulate Sailor*, 62.

O, to sail to sea in a ship!
To leave this steady, unendurable land,
To leave the tiresome sameness of the streets, the sidewalks and the houses,
To leave O you solid motionless land, and entering a ship,
To sail and sail and sail!
Walt Whitman, *Leaves of Grass* (New York, NY: Small, Maynard & Company, 1897), 221–222.

"Then just about sunrise we got for an hour an inexplicable, steady breeze right in our teeth. There was no sense in it. It fitted neither with the season of the year, nor with the secular experience of seamen as recorded in books, nor with the aspect of the sky. Only purposeful malevolence could account for it." Joseph Conrad, quoted in Tazelaar, *The Articulate Sailor*, 63.

"The sea, washing the equator and the poles, offers its perilous aid, and the power and empire that follow it. . . . Beware of me, it says, but if you can hold me, I am the key to all the lands." Ralph Waldo Emerson. http://www.quotecosmos.com/subjects/1064/Sea (August 17, 2008).

"For the army, water is a forbidding barrier, for the navy, a broad and inviting highway, and these are habits of mind engendered from earliest training." Eric Larrabee, *Commander in Chief: Franklin Delano Roosevelt, His Lieutenants and Their War* (New York, NY: Harper and Row, 1987), 336.

"It might be argued that before the days of canned foods and refrigeration, considerations of stowage space for provisions and water were not the only factors determining the time that a vessel might remain at sea. But one of the first requirements of a good seaman was a zinc-lined stomach and a relative indifference not only to sensations of smell and taste concerning food, but to those of sight as well." Bernard Brodie, *Machine Age*, 112.

"The start line is the forward edge of the forming up. . . . It must be secure and should be at right angles to the objective. It is used to help align the attacking troops with the objective." Holmes, *Acts of War*, 1.

"The major sea lines of communications used by tankers from the Middle East are also the Straits of Malacca and Singapore, the transit route for 11 million barrels a day of oil. Tankers that exceed 222,000 dwt have to divert through the Lombok Strait. The three-day detour adds US $200,000 to $300,000 per day to the voyage cost . . . . for the calendar year 1999 . . . there were 75,510 passages through the Malacca Straits by 8,678 ships." L. G. Cordner, "Maritime Terrorism: the Next 'Soft Target'?" *Defense and Foreign Affairs Daily* (December 9, 2003): 191.

"Climate and culture are obviously related in complex ways, climate being an observable and measurable artifact of culture, and considered by many to be one of the major determinants of organizational effectiveness. More recent research indicates that other cultural traits, such as involvement, consistency, adaptability, and mission orientation are positively related not only to members' perceptions of organizational effectiveness, but also to objective measures of the same. Such definitions would seem to establish from the outset that those who tinker with the culture and climate of military organizations may well be, either unknowingly or without concern, modifying the long-term effectiveness of America's armed forces." Snider, "Uninformed."

"Military genius is rooted in the ability to deal with simplicities, to preserve them amidst methodological complexity. Political genius, conversely, is the ability to deal with complexities, to reduce them to operational simplicities without recklessly disregarding their inherent intricacies." D. A. Zoll, "New Aspects of Strategy," *Strategic Review* (Fall 1973): 41.

"Since navies, alone of the armed services, operate essentially in an international environment their connection with international law has always been obvious." D. P. O'Connell, *The Influence of Law on Sea Power* (Annapolis, MD: Naval Institute Press, 1975), 1.

"The interest in global mobility seeks to avoid impediments to the deployment of forces by sea anywhere in the world. This interest is ordinarily associated with naval powers. In fact, the security of almost every state depends in some measure upon the mobility of the forces of naval powers for the maintenance of stability and security in its region." Oxman, "Territorial Temptation," 841.

"This traditional Freedom of the Seas Regime had prevailed since the early years of modern civilization and reflected the simple fact that the economic value of the oceans was based almost entirely on their utility as a medium for trade." S.W. Haines, "The Maritime Domain: Security, Law Enforcement, and Control Requirements in the Offshore Zones," *Naval Forces* 4 (1989): 16.

"Freedom of the seas in peace and sea supremacy in war are basic to the existence of the United States as a world power. They are keystones of United States foreign policy and are essential to the Free World cooperation which it nurtures." Arleigh Burke, "Origins of United States Navy Doctrine," *Attachment to Letter to All U.S. Naval Officers, Officer Candidates, Midshipmen and Cadets* (11 April 1960), 11.

"[S]ome 90 percent of them [the world's ocean fisheries] have been placed under the largely discretionary control of coastal states by virtue of the regime of the EEZ." Oxman, "Territorial Temptation," 841.

"95% of the experts believe that more States will claim the right to exercise jurisdiction and control over military activities in their exclusive economic zones (EEZ) by 2020." *Final Report on the Legal Experts' Workshop on the Future Global Legal Order* (Newport, RI: Naval War College, 2006), iii.

"'The whole art of war consists of getting at what is on the other side of the hill,' said the Duke of Wellington, conqueror of Napoleon at Waterloo. In the murky kind of fight that marks modern warfare against terrorists and guerrillas, knowing what's on the other side of the hill—or inside a building—takes on a whole new urgency and meaning." John Barry and Evan Thomas, "Up In The Sky, An Unblinking Eye," *Newsweek* (June 9, 2008).

"There are no smooth waters at sea." William Robinson, quoted in James Tazelaar, *Articulate Sailor*, 67.

"Confronting a storm is like fighting God. All the powers in the universe seem to be against you and, in an extraordinary way, your irrelevance is at the same time both humbling and exalting." Franciose LaGrange, http://www.pbase.com/image/33827233 (August 17, 2008).

"I was once more alone with myself in the realization that I was on the mighty sea and in the hands of the elements." Joshua Slocum, *Sailing Alone Around the World*. http://www.scribd.com/doc/882667/Sailing-Alone-Around-The-World-by-Joshua-Slocum (August 17, 2008).

"He that will not sail till all dangers are over must never put to sea." Thomas Fuller, *Gnomologia* (London: Printed for B. Barker, 1732), 97.

"The sea hates a coward." Eugene O'Neill. http://thinkexist.com/quotation/the_sea_hates_a/165884.html (August 17, 2008).

"There is nothing more enticing, disenchanting, and enslaving than the life at sea." Joseph Conrad. http://thinkexist.com/quotation/there_is_nothing_more_enticing-disenchanting-and/255833.html (August 17, 2008).

"If you want to build a ship, don't herd people together to collect wood and don't assign them tasks and work, but rather teach them to long for the endless immensity of the sea." Antoine de Saint Exupery. http://www.elise.com/quotes/a/antoine_de_saintexupery_if_you_want_to_build_a_ship.php (August 17, 2008).

"The wind and the waves in a naval battle are always on the side of the ablest navigator." Edmund Gibbon and Robert Christy, *Proverbs, Maxims, and Phrases of All Ages*, vol. 1 (New York, NY: G. P. Putnam's Sons, 1893), 57.

"My imagination refuses to see any sort of submarine doing anything but suffocate its crew and founder." H. G. Wells, *Anticipations: Of the Reaction of Mechanical and Scientific Progress upon Human Life and Thought,* 1906, quoted in Gordon G. Feir, *H. G. Wells at the End of His Tether* (n.p.: iUniverse, 2005), 175.

"The language of the sea is the vernacular of a hard life." Gershom Bradford, *A Glossary of Sea Terms* (New York, NY: Dodd, Mead & Co,. 1943).

"Ports are no good—ships rot, men go to the devil." Joseph Conrad. http://www.accidentalcruiser.com/journals/journal070611.html (August 17, 2008).

"By nature the land is almost all obstacle, the sea almost all open plain." Mahan, *Mahan on Naval Strategy,* 136.

"While there are fewer obstacles they are truly impassable to ships and ships cannot force them, but must go around." Mahan, *Mahan on Naval Strategy,* 139.

"The battlefields of the sea bear, of course, no physical trace of the events that transpired in those places; wind and water wipe the debris from the surface in a few days, even hours, and the depths engulf the ships and men that fell victim to the action. Land battlefields are marked more lastingly. The soldier's space leaves scars that may persist for a hundred years, as those of the American Civil War still do." John Keegan, *The Price of Admiralty: The Evolution of Naval Warfare* (New York, NY: Viking, 1989), 10.

"You may look at the map and see flags stuck in at different points and consider that the results will be certain, but when you get out on the sea with its vast distances, its storms and mists, and with night coming on, and all the uncertainties which exist, you cannot possibly expect that the kind of conditions which would be appropriate to the movements of armies have any application to the haphazard conditions of war at sea." Winston Churchill to the House of Commons, October 11, 1940. Quoted in Winston Churchill, *The Second World War,* vol. 1 (Boston, MA: Houghton Mifflin, 1986), 542.

"There are few, if any, characteristics of the utterances which I from time to time hear, or read, on the subject of actual warfare, which impress me more strongly than the constantly recurring tendency to reject any solution of a problem which does not wholly eliminate the element of doubt, of uncertainty, or risk. Instead of frankly recognizing that almost all warlike undertakings present at best but a choice of difficulties—that absolute certainty is unattainable—that the 'art' consists, not in stacking the cards, but as Napoleon phrased it, in getting the most of the chances on your side—that some risk, not merely of death but of failure, must be undergone—instead of this, people wish so to arrange their programme as to have a perfectly sure thing of it; and when some critic points out, as can so easily be done, that this may happen or that may happen, and it is seen undeniably that it may, then the plan stands condemned. 'War,' said Napoleon again, 'cannot be made without running risks and it is because my admirals have found out that it can, everything attempted by them has failed.' " Alfred Thayer Mahan, "Blockade In Relation To Naval Strategy," U.S. Naval Institute *Proceedings* (November 1895): 852–853.

Chapter 4: Strategies for the Employment of Naval Forces

"Seapower is our heritage. . . . Seapower is not merely a fleet of ships and planes. Seapower reaches into every phase of our national life. It means domination of the sea and the air above it. In order to maintain such power the fruits of our farms, the minerals of our land and the products of our factories must flow into the sinews of maritime strength." Nimitz, undated post—World War II quote. Courtesy Naval Historical Center, Washington, D.C.

"Operationally and tactically, naval power cannot offset ground forces save at the margin. But strategically, such power can limit the consequences of adverse events on land and, more important can—and repeatedly has—enable a maritime combatant to protract a war in time, extend it in geography, and assemble a coalition able to field a superior landward fighting instrument." Gray, *War, Peace, and Victory*, 76.

"You will be governed by the principle of calculated risk, which you shall interpret to mean the avoidance of exposure of your force to attack by superior enemy forces without good prospect of inflicting . . . greater damage on the enemy." Chester A. Nimitz, prior to the Battle of Midway. http://www.cv6.org/1942/midway/default.htm (August 18, 2008).

"Dulles . . . pressed hard for intervention [in Indochina], but . . . he surrounded it with so many conditions that were unlikely to be met in the circumstances of the time that one might be forgiven for thinking that he never really wanted to make a military commitment. 'Dulles,' said Georges Bidault, who was the French Foreign Minister at the time, 'was always talking of 'calculated risks,' which in practice most often meant that he calculated a great deal and risked nothing.' " Bidault, quoted in Theodore Draper, *Abuse of Power* (New York, NY: The Viking Press, 1967), 34.

"But strategy is not a magical elixir that properly fortified society can use to accomplish its ends. That U.S. strategic performance in Vietnam was so lamentable does not mean that a better strategy could have produced success. Unsuitable policy choices can frustrate strategy just as much as operational incompetence or tactical ineptitude." Colin S. Gray, "Strategy in the Nuclear Age: The United States, 1945–1991," in Williamson Murray, MacGregor Knox, and Alvin Bernstein, *The Making of Strategy: Rulers, States and War* (New York, NY: Cambridge University Press, 1994), 581.

"There is no zero-risk situation in war. The willingness to run a calculated risk and to absorb some damage is essential. In sum, heroes run risks. Smart heroes calculate the risks and take steps to shift the odds more in their favor. Those who avoid risks stay home." H. Dwight Lyons Jr., Eleanor A. Baker, Sabrina R. Edlow, and David A. Perin, *The Mine Threat: Show Stoppers or Speed Bumps?* (Alexandria, VA: Center for Naval Analyses, 1993), 19.

"Seapower means flexibility and fluidity in the use of other military power. The seas enable us to apply relatively small forces and yet achieve superiority in critical areas of our own choosing." James V. Forrestal, secretary of the Navy, speech before the Economic Club of New York, Hotel Astor, New York (February 25, 1947). http://www.ibiblio.org/pub/academic/history/marshall/military/navy/USN_wwii_quotes.txt (August 18, 2008).

"Biddlecomb felt his thoughts wandering, sailing out across the black, rain-swept harbor. To most it was a frightening and dangerous place, but to him it was a sanctuary. On the water one could not become trapped as easily as one could on land. On the water there was always someplace to run, and once clear of the constricting land one could circle the globe on contiguous seas." James L. Nelson, *Lords of the Ocean* (New York, NY: Pocket Books, 1999), 47.

"Nelson knew exactly the risks he ran and accurately allowed for them. He had clear knowledge, from long-considered fighting experiences, how long his ships could endure the temporary gunnery disadvantage necessary in order to gain the dominant tactical position he aimed at for a great victory. . . . We had to buy that experience [at Jutland], for our weapons were untried. The risks could not be measured without that experience. . . . Dreadnoughts had never engaged, modern massed destroyer attack had never taken place." Ernle Chatfield, Admiral Beatty's staff commander of the battlecruiser fleet at the Battle of Jutland, quoted in John Keegan, *The Price of Admiralty*, 154.

"The key to anti-terrorism strategy is to eliminate safe havens." Henry Kissinger, "Crucial Phase Of America's Anti-Terrorism Strategy Yet To Come," *San Diego Union-Tribune*, October 28, 2001.

"In contrast to the land, the sea is a medium for movement. It cannot be occupied and fortified." Gray, *War, Peace, and Victory*, 63.

"The supreme test of the naval strategist is the depth of his comprehension of the intimate relation between sea power and land power, and of the truth that basically all effort afloat should be directed at an effect ashore." Dudley W. Knox, *The Naval Genius of George Washington* (Boston, MA: Houghton Mifflin, 1932), 5.

"[T]he Soviet Navy does not build sea-control weapons systems, but sea-denial weapons systems such as submarines and raiding cruisers supplemented by land-based aircraft and minelaying. The history of warfare teaches us that such attempts at denial are always the expedient adopted by a power which is too weak in capital ships to fight for control of the sea. History also teaches us that such attempts have uniformly ended in failure, though the idea continues to have a seemingly irresistible attraction for the statesmen of countries more habituated to land than to sea warfare, such as Russia or Germany, in World Wars I and II." George Fielding Eliot, *Victory without War 1958–1961* (Annapolis, MD: United States Naval Institute, 1958), 108.

"Never fight the United States without nuclear weapons." Thomas Mahnken, "Deny U.S. Access?" Naval Institute *Proceedings* (September 1998): 38.

"Appreciating America's great superiority in naval assets, the Mexicans did not contest command of the sea. Instead, they adopted the traditional strategy of weaker naval powers: they relied on privateers to raid enemy commerce and they defended the coastline with fortifications—essentially the same strategy used by the United States against Great Britain in two wars and the strategy the Confederacy would adopt in 1861. For its part, the United States relied on the strategy traditionally practiced by the British in their wars against nonnaval powers: blockading the coast, suppressing commerce raiders, and operating against the shore—a strategy that foreshadowed Union operations against the Confederacy." Craig L. Symonds, *Confederate Admiral: The Life and Wars of Franklin Buchanan* (Annapolis, MD: Naval Institute Press, 1999), 8

"Contending that American strategy should not be based on total 'strategic' bombing as a course of action, Radford reasoned that to spend millions on huge land-based aircraft designed for high-level 'carpet bombing' was militarily ineffective, morally wrong, and politically disastrous. On the other hand, the naval carrier task force, with its inherent mobility and flexibility, could deliver precision attacks on designated targets and achieve far more effective results." Quoted from Paul B. Ryan, *First Line of Defense: The U.S. Navy Since 1945* (Stanford, CA: Hoover Institution Press, 1981), 12.

"There is probably more nonsense talked and written at every level of classification from Top Secret to the Washington Post about anti-submarine warfare than any other military subject." Richard Sharpe, *Jane's Fighting Ships 1988–1989*, 91st ed. (London: Jane's Publishing, 1988), 105.

"In spite of its performance as a commerce destroyer, the submarine neither supplanted the battle fleet nor drove it from the seas. Its success against warships was in more confined waters where it could lie in ambush. During the war, most big ships were torpedoed at the approaches to naval bases, such as Heligoland Bight, or in the area of shore bombardment, such as the Dardanelles. No first line capital ship was sunk by a submarine [in World War I]." Karl Lautenschlager, "Technology and the Evolution of Naval Warfare," *International Security* 8, no. 2 (Autumn 1983): 20.

"To a naval officer it is . . . foreign to his thinking that we would restrict our attacks to a submarine at sea." Claude V. Ricketts, quoted in Richard Hegmann, "Reconsidering the Evolution of the US Maritime Strategy 1955-1965," *Journal of Strategic Studies* 14, no. 3 (September 1991): 311.

"At the first meeting of the North Atlantic Ocean Regional Planning Group (what would become SACLANT) held in Washington in the fall of 1949, American and British admirals adopted 'attack at source' as NATO doctrine, and discussed an offensive against Soviet bases in the Kola." Michael Palmer, *Undersea Warfare and Maritime Strategy: The American Experience.* Paper delivered at the *Undersea Dimension of Maritime Strategy Conference*, Halifax, Nova Scotia (June 22, 1989), 21.

"One of the standard principles in naval strategy has been what is called the 'fleet in being'—the fact that all sea movements in a given area are affected by the mere presence of a strong naval force in that area, and that that influence persists so long as that force is present as a potential threat, even though it may never be engaged. There is no similar war-time principle applicable to air forces. . . . the tonnage, armour, speed, and armament of naval ships are positive and calculable factors which can be decisive in action, and there is no doubt that the 'fleet in being' continued during the late war to have a great influence on strategy an on operation at sea and in the air. In this game the most powerful ships have so far held the trumps. Although no fleet action took place, the mere presence in German waters of a fleet, small in numbers though it was, involved us in a continuous commitment as regards the maintenance, development and dispositions of our own fleet. It involved a considerable commitment for our air forces as well . . . In the Pacific war, although most of the actions were between carrier-borne aircraft and ships and not directly ship to ship, the existence of a Japanese battle fleet led to the creation of a bigger and better American battle fleet, which held the ring while the carriers operated." Arthur William Tedder, *Air Power in War* (London: Hodden and Stoughton, 1948), 84–85.

"The essential difference is that war is not an exercise of will directed at inanimate matter, as is the case with the mechanical arts. . . . In war, the will is directed at an animate object that reacts. It must be obvious that the intellectual codification used in the arts and sciences is inappropriate to such an activity." Carl von Clausewitz, *On War*, bk. 2, Michael Howard and Peter Paret, eds. (Princeton, NJ: Princeton University Press), 149.

"In this century only the Japanese Navy has been able to apply in war the strategy it had devised in peace. Other navies had to discard their peacetime notions, because new weapons (submarines in World War I, aircraft in World War II) assumed unexpected importance, or because the enemy achieved strategic surprise." Sir James Cable, "More Crucial Now Than Ever," Naval Institute *Proceedings* (May 1992): 86.

"In the future, as in the past, the key to victory and to the freedom of this country will be in the control of the seas and of the skies above them. Attacks upon us or attacks by us must cross on, over, or under the sea. That fact is an accident of geography which you can confirm by any map. No enemy can reach us without coming across the sea. Therefore, control of the ocean and of the air over it is the key to our own security. . . . The control of the sea and the air above it is the mission of the United States Navy." James Forrestal, quoted in Kenneth J. Hagan *This People's Navy: The Making of American Sea Power* (New York, NY: The Free Press, 1991), 335.

"In fact, of course, 'sea lanes' are mathematical abstractions, lines drawn upon charts; it is ships which must be defended, not featureless stretches of water." E. Cameron Williams, "Sail Together or Sink Separately: The Convoy and Risk Analysis," *Defense Transportation Journal* (April 1987): 15.

"I believe we err in advancing the proposition that 'control of the sea' is an end in itself. It is the exploitation of this control that is important." Vice Adm. Richard L. Conolly, quoted in Geoffrey Till, *Maritime Strategy and the Nuclear Age* (New York, NY: St. Martin's Press, 1982), 192.

"One of the most effective ways to interdict oceanic (sea) shipping is the destruction of the primary elements of maritime transport, in particular, the port management infrastructure. Another, no less important direction of this warfare can be depriving the enemy of the capability to replace losses of transport tonnage and convoy escort vessels. Both of these methods inevitably result in massive, including nuclear, strikes on naval bases, ports, shipbuilding centers and other important military and economic coastal targets." Admiral of the Fleet V. Chernavin, "The Struggle for the Sea Lanes of Communication: Lessons of Wars and the Modern Era," *Morskoy Sbornik,* 2 (February 1990). Translated in JPRS-UMA-90-007 (23 March 1990): 67.

"The successful strategist is the one who controls the nature and the placement and the timing and the weight of the centers of gravity of war, and who exploits the resulting control of the pattern of war toward his own ends." Wylie, *Military Strategy*, 91.

"From the days when humans first began to use the seas, the great lesson of history is that the enemy who is confined to a land strategy is in the end defeated." Field

Marshal Viscount Montgomery of Alamein, quoted in Geoffrey Till, ed., *Seapower in Theory and Practice* (London: Frank Cass, 1994), 18.

"In the world of strategists, as opposed to that of tacticians, there is simply much more scope for error. Finally, it is critical to flag an underrecognized source of friction, the will, skill, and means of an intelligent and malevolent enemy." Gray, "Why Strategy Is Difficult," 10.

Chapter 5: Expeditionary

"The whole principle of naval fighting is to be free to go anywhere with every damned thing the Navy possesses." Sir John Fisher, *Memories*, 1919. Quoted in Heinl, *Dictionary*, 207.

"The Tripolitan war provided a plain object lesson in naval strategy, by showing the decisive advantage of seizing command of the home waters of a maritime enemy. Only after the American Navy had established such superiority as enabled it to blockade the coast of Tripoli, was it possible to force the war to a successful conclusion." Harold and Margaret Sprout, *The Rise of American Naval Power 1776–1918* (Annapolis, MD: Naval Institute Press, 1939), 57.

"Did you know the war on the Barbary pirates was not declared? All Jefferson got from Congress was the authorization to build some ships and an AUMF [Authorization to Use Military Force] which reads like the AUMF against Iraq almost word for word. Plus the war went on for over a decade and extended into Madison's term." http://powerandcontrol.blogspot.com/2008_07_01_archive.html (October 24, 2008).

"Weakness provokes insult and injury, while a condition to punish often prevents them. This reasoning leads to the necessity of some naval force; that being the only weapon by which we can reach an enemy. I think it to our interest to punish the first insult; because an insult unpunished is the parent of many others. We are not at this moment in a condition to do it, but we should put ourselves into it as soon as possible." Thomas Jefferson to John Jay, August 23, 1785, *The Letters of Thomas Jefferson*. http://www.yale.edu/lawweb/avalon/jefflett/let32.htm (August 19, 2008).

"If naval officers believed they had prior claim to the disposition of an incident, if they felt themselves more capable of handling the problem than the local State Department authorities, or if they simply disapproved of the local U.S. consul or commercial agent, then they might be inclined to take matters into their own hands. Many officers would take to heart Assistant Navy Secretary Theodore Roosevelt's curious remark to the Naval War College in 1897: 'The diplomat is the servant, not the master, of the soldier.' " Karsten, *Naval Aristocracy*, 173.

"We are a global force. We are about being around the world, around the clock, anywhere and anytime. We do this with an awesome global naval force. We operate from the maritime domain. We are about taking the sovereignty of the United States of America to the far corners of this earth, and we do it in places where we don't need a permission slip from anybody." Adm. Vern Clark, *Remarks at the United States Naval Academy Commencement*, Annapolis, Maryland, May 24, 2002.

"This rich set of data allows Thompson to compare the evolution of great sea powers, sometimes centuries apart. While the historian might balk at this approach at first glance, he or she must keep in mind how few of the parameters of sea power have fundamentally changed over the centuries. The physical geography has remained the same for tens of thousands of years. Moreover, twentieth-century battles almost invariably have taken place in the same locations as those of centuries prior. . . . Even the building blocks of sea power—ships and fleets—retained many of their essential characteristics between Actium and Dogger Bank." Mark R. Shulman, "Hitting the Target: Perspectives on Doing Naval History," in John B. Hattendorf, ed., *Doing Naval History: Essays Toward Improvement*, (Newport, RI: Naval War College Press, 1995), 154–155.

"The threat to the coasts during at least the early days of World War II was real—second in importance in U.S. history only to the threat during the War of 1812. The Japanese attacked and took several U.S. island possessions in the far and mid-Pacific, despite valiant but underresourced U.S. Army, Navy, and Marine Corps inshore, air, and ground defense efforts. The Japanese also conducted a few nuisance submarine shore bombardments and air raids on the West Coast. On the East Coast, however, in the spring of 1942, long-range German submarines sank a large amount of American coastal shipping, laid minefields, and landed saboteurs in New York and Florida. Nevertheless, the U.S. Navy's vision stayed forward, focused first on the need to carry the war across the Pacific to the Japanese and, second to assist in fighting Germans and Italians across the Atlantic and in the Mediterranean." Swartz, *Forward . . . From the Start*, 9.

"Ability to use the ocean spaces in war and to deny their use to opponents assures the strategic initiative, even in adversity." Arleigh Burke, "Origins of United States Navy Doctrine," Attachment to *Letter to All U.S. Naval Officers, Officer Candidates, Midshipmen and Cadets* (11 April 1960): 11.

"Far to the northward were to be seen white topsails in Quiberon Bay—Hornblower, from the deck of the Sophia, saw signals pass back and forth from the Indefatigable as she reported her arrival to the senior officer of the main expedition there. It was a proof of the mobility and ubiquity of naval power that it could take advantage of the configuration of the land so that two blows could be struck almost in sight of each other from the sea yet separated by forty miles of roads on land." C. S. Forester, *Mr. Midshipman Hornblower* (New York, NY: Pinnacle Books, 1948), 129–30.

"Of highest importance, a sea-based [ABM] system deployed close to the USSR would greatly degrade the value to the USSR of both MIRV's and decoys, for interception of ICBM's on the ascending leg of the trajectory permits intercepting the missile before the multiple warheads are launched." Paul C. Davis, "Sentinel and the Future of SABMIS," *Military Review* (March 1968), 196.

"Von der Goltz had made clear before the turn of the century: No matter how fine an army, no matter how well they were led, no matter how bravely they might fight—the best result obtainable on the strategic defensive is stalemate." Harry G. Summers, *On Strategy II* (New York, NY: Dell Publishers, 1992), 185.

"[I]n giving up the offensive the navy gives up its proper sphere, which is also the most effective." Alfred Thayer Mahan, *Naval Strategy, Compared and Contrasted with the Principles and Practice of Military Operations on Land* (Boston, MA: Little, Brown, and Company, 1919; first pub. 1911), 153.

"The advantage of sea-power used offensively is that when a fleet sails no one can be sure where it is going to strike." Winston Churchill, *Their Finest Hour*, 1949. Quoted in Heinl, *Dictionary*, 208.

"Cease firing, but if any enemy planes appear shoot them down in friendly fashion." Halsey in Taylor, *Language of World War II,* and *Newsweek*, August 27, 1945. Courtesy Naval Historical Center, Washington, D.C.

Put the missiles out to sea
Where the real estate is free
And they're far away from me.
— Anonymous

"The German blitzkrieg through France, often regarded as the pinnacle of swift, nearly battle-free operational maneuver, moved just two hundred miles in seven days, or thirty miles per day; the French army never recovered. Robert L. Helmbold's definitive study of army movement rates suggests that against light opposition an average advance of a hundred miles a week is more typical. A fleet at sea moves a greater weight of combat potential more than an order of magnitude faster than the blitzkrieg—2,500 miles in seven days." Wayne P. Hughes Jr, "Naval Maneuver Warfare," *Naval War College Review* 50, no. 3 (Summer 1997): 28.

"The U.S. has already decreased its overseas presence from 115 major bases in 1956 to 27 in 1995." Office of Naval Intelligence, *Challenges to Naval Expeditionary Warfare* (n.p.: Office of Naval Intelligence, March 1997), 7.

' "If I am in command when war breaks out I shall issue my orders: The essence of war is violence. Moderation in war is imbecility. Hit first, hit hard, and hit everywhere." Admiral Sir John Fisher, April 25, 1912, letter to Lord Esher, quoted in Francis Whiting Halsey, *The Literary Digest History of the World War* (New York, NY: Funk & Wagnalls Company, 1920), 150.

"Adaptive learning requires high rates of feedback from the environment to drive continuous change and modification. . . . The system does not wait for specific elements of information before it acts—it can perform quite well by acting on the anticipation itself. Adaptive response allows a system to respond smoothly to changes in information conditions, avoid 'knee-jerk,' reactive responses and more likely survive shocks to the environment." Jeff Cares, *Distributed Networked Operations: The Foundations of Network Centric Warfare* (New York, NY: iUniverse, Inc., 2005), 14.

"War, in common with sport, has the characteristic that what worked yesterday may not work tomorrow, precisely because it worked yesterday." Douglas A. MacGregor, *Breaking the Phalanx: A New Design for Landpower in the Twenty-first Century* (Westport, CT: Praeger, 1997), 94.

"The non-linear and fluid operating environments that will characterize future battlefields necessitate renewed emphasis on both adaptive planning and dynamic military operations. In the Gulf War, for example, 20 percent of targets were selected after aircraft launch, whereas over Kosovo 43 percent of targets were selected once the aircraft were airborne. In Afghanistan, 80 percent of the carrier-based sorties were launched without designated targets. Operation Iraqi Freedom statistics are likely to be consistent with this trend." Robert P. Haffa Jr., and Robert E. Mullins, "Trends in America's Post-Cold War Military Conflicts: The Implications for Sea Power," *Sea Power* (July 2003): 14.

"The contestant who controls the pattern of the war has an inestimable advantage. He can, in great measure, call the tune and make the opponent dance to it." Wylie, *Military Strategy*, 159.

"Admirals Spruance and Fletcher [at Midway], entirely on their own, sent almost the entire American air arm after the Japanese at the first opportunity. Such actions by Spruance and Fletcher may have been precipitous, but they were grounded in the wisdom that in carrier war the first strike is often the most critical, since it can wipe out the enemy's ability to retaliate and can obliterate the platform itself for hundreds of planes in the air." Hanson, *Carnage and Culture,* 379.

"Where are you bound?" Algerine warship captain to Commodore Stephen Decatur passing in the Mediterranean in summer 1815. "Where I please!" answered Decatur. London, *Victory in Tripoli*, 241.

"Victory smiles on those who anticipate the changes in the character of war, not upon those who wait to adapt themselves after changes occur." Giulio Douhet, *The Command of the Air* (New York, NY: Cowardd-McCann, 1942), 30.

Chapter 6: Technology and Systems

Link to Naval Weapons, Technology, and Reunions: http://www.navweaps.com. (August 20, 2008)

"The U.S. Navy in the postwar era may well have actively promoted and managed a wider array of technology than any other single institution in a comparable period in world history." David A. Rosenberg, in N. A. M. Rodger, ed., *Naval Power in the Twentieth Century* (Annapolis, MD: Naval Institute Press, 1994), 244.

"A second reason for the self-referencing nature of navies is their developing role as symbols of the states to which they belong. In this context Nicholas Rodger shows the way by his statement that warships involve the most advanced and complex technologies of their eras. . . . Warships are indeed, in the words of Robert O'Connell, 'sacred vessels' whose expendability not infrequently reflects moral as well as operational considerations. The British Army could suffer 60,000 casualties in a single day, but Jellicoe could have lost Britain's war in an afternoon. Given that context it is understandable that much naval writing is not only national in focus, but reflects particular national mythologies in ways military history does not." Dennis E. Showalter, "Toward a 'New' Naval History." in Hattendorf, *Doing Naval History,* 13

"Their [U.S. fighting ships] utility in the War of 1812 literally stunned the Royal Navy. The heroic tradition they inspired bonded the young U.S. Navy to the giant frigate image as its badge of identity and national uniqueness." Michael E. Vlahos, "The Making of an American Naval Tradition (1795–1887)," in Firebaugh, *Naval Engineering,* 5.

"Most of the written history of wayfinding concerns the invention or adaptation of tools to support nautical exploration. The limited availability of landmarks and seamarks combined with the high cost of getting lost to provide a powerful incentive to be inventive. Consider the following solutions employed by sailors over the centuries: Lighthouse . . . Compass . . . Chip log . . . Sextant . . . Chronometer . . . Maps and Charts." Morville, *Ambient Findability,* 47.

"In war, as in life generally, all parts of the whole are interconnected and thus the effects produced, however small their cause, must influence all subsequent military operations and modify their final outcome to some degree, however slight. In the same way, every means must influence even the ultimate purpose. Interconnectedness and context, interaction, chance, complexity, indistinct boundaries, feedback effects and so on, all leading to analytical unpredictability." Alan Beyerchen, "Clausewitz, Nonlinearity, and the Unpredictability of War," *International Security* 17, no. 3 (Winter 1992–1993): 82.

"Technique is as important as tools. Put those two things together and you have a technological system. That system can be as simple as going to a well, lowering a bucket, pulling up water, and putting it on a wagon to carry it somewhere; or it can be as complex as a locomotive. They're all technological systems. Maury Klein, "The Technological Revolution," Foreign Policy Research Institute, *Footnotes: The Newsletter of FPRI's Wachman Center* 13, no. 18 (July 2008).

"It has long been appreciated that the critical weaknesses in French naval power in the age of sail lay in personnel and doctrine rather than technology; it is much less well appreciated that the same point applied to the Imperial German Navy also." Gray, *War, Peace, and Victory*, 393.

"Because of new surveillance measures, you could have whole zones of the ocean where you are unable to operate safely on the surface, Donald Henry, special assistant to the director of the Pentagon's Office of Net Assessment, told me. . . . The faster technology progresses, the less likely people will play by our rules." Robert D. Kaplan, "America's Elegant Decline," *Atlantic Monthly*, October 3, 2007. http://www.ocnus.net/artman2/publish/Defence_Arms_13/America_s_Elegant_Decline.shtml (August 20, 2008).

Secretary McNamara testified: "I have seen no arguments that would justify in any way the additional costs associated with a nuclear-powered carrier." U.S. Congress, House, *Hearings* Defense Procurement Authorization FY 1963, Paper No. 44, 87/2, 1962, p 3258–3259.

"No Islamic nation could have flown to the moon or invented the Internet, simply because for a millennium the culture has suppressed the curiosity necessary for such a venture." Mark Steyn, *America Alone: The End of the World as We Know It* (Washington, D.C.: Regnery Publishing, Inc., 2006), 17.

"Good decision makers in complex environments recognize that it is a fool's errand to merely increase precision. They know that the command and control process should be directed at discovering the motives, dynamics and long-term evolutionary behaviors of competitors in the environment. These decision makers try to attain more than a reactive response to information on the battlefield; they attempt to create a deeper mental model of the environment to anticipate how competition might unfold." Cares, *Distributed Networked Operations*, 16.

"Systems analysts, however, have been telling us for years that in organizations the links between cause and effect can be complex, distant in time and space, and very difficult to detect. . . . Basically, the dynamics of even simple nonlinear feedback systems are so complex that the links between cause and effect are lost in the detail of what happens. Tiny changes can escalate to have massive consequences; virtuous and vicious circles are generated. It is therefore totally impossible to predict the specific long-term future of such a system; that future is truly unknowable." Ralph D. Stacey, *Managing the Unknowable: Strategic Boundaries between Order and Chaos in Organizations* (San Francisco: Jossey-Bass Publishers, 1992), 11.

"The heart of the matter is that the system's variables cannot be effectively isolated from each other or from their context; linearization is not possible, because dynamic interaction is one of the system's defining characteristics." Beyerchen, "Clausewitz, Nonlinearity," 66.

"We decentralize and capitalize on the capabilities of our individual people rather than centralize and make automatons of them. This builds that essential pride of service and sense of accomplishment." David A. Rosenberg, "Arleigh Burke," in Robert William Love Jr., ed., *The Chiefs of Naval Operations* (Annapolis, MD: Naval Institute Press, 1980), 287.

"In the Turkish strait in 1996, the nine pro-Chechen gunmen who hijacked a Turkish ferry and held 255 passengers hostage for three days had first considered the possibility of sabotaging one of the two suspension bridges with explosives to bring down the bridge and close shipping traffic." Tony Corn, "The Revolution in Transatlantic Affairs," August 21, 2007. http://www.realclearpolitics.com/articles/2007/08/the_revolution_in_transatlanti.html (August 20, 2008).

"In order to appreciate fully a general's capacities, the reader should not know more at any point than the commander himself knew at the time." Jac Weller, *Wellington in the Peninsula, 1808–1814* (London, 1992), xiv, quoted in Colin S. Gray, *Modern Strategy* (Oxford: Oxford University Press, 1999), 305.

"Fortunately, the U.S. Navy has been following a path that elevates the information and communication dimensions of war to high importance. For, at sea, to be located is to be come immediately vulnerable to destruction. In fact, naval war may already be arriving at a doctrine that looks a lot like cyberwar." John Arquilla and

David Ronfeldt, "Cyberwar & Netwar; Warfare between Networks: CYBERWAR IS COMING," *Comparative Strategy* 12, no. 2, in John Arquilla and David Ronfeld, *In Athena's Camp: Preparing for Conflict in the Information Age* (Santa Monica, CA: RAND, 1997), 48.

"[Clausewitz] wrote that 'since all information and assumptions are open to doubt, and with chance at work everywhere, the commander continually finds that things are not as he expected,' and that the inflow of new information only made him 'more, not less uncertain.' " Sumida, *Inventing Grand Strategy,* 112.

"Like Nelson, Burke deduced that there was 'no time in battle to give orders' and chose instead to rely on personal doctrine and the initiative of his subordinates. . . . Commanders served by today's technology receive much more data, but at the same time the increasing complexity of warfare leaves just as much if not more information beyond their grasp. Simply put, uncertainty is not a measurable amount of unknown information; rather, it is part of the human condition, a reality of life with which our minds grapple every day. Fortunately, our thought processes work in such a fashion that we are able to make sensible decisions without complete information. We spend our entire lives functioning within a fog of uncertainty. . . . Uncertainty is also a precondition of warfare. In the absence of uncertainty, there would be no conflict. . . . Over the centuries navies, far more than armies, have employed decentralized approaches. Until the late nineteenth century admiralties had no choice at the strategic and operational levels of war but to trust to the initiative of subordinates operating on distant stations." Palmer, *Command at Sea*, 320–321.

"[H]istory also demonstrates that decentralization is not suitable for all navies at all times. . . . Moreover, in a modern world shadowed by the threat of catastrophic nuclear escalation, restrictive rules of engagement and micromanagement may well be justified. Nor does decentralization guarantee succes. . . . And yet the victories of the decentralizers over the past several centuries suggest that in battle commanders blessed with capable, well-trained and well-indoctrinated men and women ought not to expect the latest communicative technology to ensure triumph, but should rely instead, as much as reasonably possible, on the talents, judgment, and initiative of their subordinates. Despite major advances in technology, wars are still fought by people, and finding the best means to exploit their talents remains central to the realization of victory." Palmer, *Command at Sea*, 322.

"Information technology may be transforming the U.S. Navy from its WW II Pacific ideal to its own Jutland with ship captains robbed of initiative and independent command." Robert D. Worley, *Shaping U.S. Military Forces: Revolution or Relevance After the Cold War* (Arlington, VA: Lulu, 2005), 216.

"Radar, the gas turbine, the airplane, nuclear propulsion and satellite communications [and navigation] have revolutionized the world in which he lives. But the sea is still there." Winnefeld, "Why Sailors Are Different," 68.

"Technological improvements often help drive Navy homeland defense systems into better uses forward. As underwater explosives, turreted gun mounts, torpedo boats, submarines, aircraft, inshore undersea warfare equipment, SOSUS, and other systems developed longer range and more robustness, they moved from being harbor and coastal defense systems to forward deployable and deployed systems." Swartz, *Forward From the Start,* 13.

"This is the paradox of Ambient Findability. As information volume increases, our ability to find any particular item decreases." Morville, *Ambient Findability*, 86.

"A preoccupation with fighting only short, high-tech, low casualty wars is virtually a tenet of U.S. national military strategy. It is a weakness, not a strength." Grant T. Hammond, "Paradoxes of War," *Joint Force Quarterly* (Spring 1994): 13.

Chapter 7. The Maritime Strategy of the 1980s

"What the distant future of the atomic research will bring to the fleet which we honor today, no one can foretell. But the fundamental mission of the Navy has not changed. Control of our sea approaches and of the skies above them is still the key to our freedom and to our ability to help enforce the peace of the world. No enemy will every strike us directly except across the sea. We cannot reach out to help stop and defeat an aggressor without crossing the sea. . . ." President Truman's address on Foreign Policy at the Navy Day Celebration, New York City, October 27, 1945. Courtesy Naval Historical Center, Washington, D.C.

"The tasks which can be given to armed forces can be reduced to the following: conquer territory or deny territory to the enemy. Destroy the enemy forces or wear them down. Act with rapidity or gain time. " Andre Beaufre, *An Introduction to Strategy With Particular Reference to Problems of Defense, Politics, Economics, and*

Diplomacy in the Nuclear Age (New York, NY: Frederick A. Praeger, Publishers, 1965), 68.

"Institute rigorous, continuous examination of enemy capabilities and potentialities, thereby getting the utmost value of information of the enemy and enabling our forces to be used with the greatest effectiveness. It is particularly important to comprehend the enemy point of view in all aspects." Fleet Admiral Ernest J. King, U.S. Navy (Ret.), Department of the Navy, *Naval Intelligence* NDP-2, September,1994.

"Freedom to use the seas is our Nation's lifeblood. For that reason our Navy is designed to keep the sea lanes open worldwide. . . . Maritime superiority for us is a necessity. We must be able in time of emergency to venture in harm's way, controlling air, surface, and sub-surface areas to assure access to all the oceans of the world." President Ronald Reagan, excerpts from the text of remarks by the president at the recommissioning of the USS *New Jersey*, Long Beach, CA, December 28, 1982. The White House, Office of the Press Secretary, December 29, 1982.

"Today, our strategy must be based on confining Russia to a land strategy, by retaining control of the seas in our hands and by preventing Russia interfering with our use of the air flank. Any other strategy is useless." Field Marshal Sir Bernard Montgomery in a 1958 lecture. Quoted in Gray, *War, Peace, and Victory*, 67.

"Beginning in 1965, the navy was instructed not to ask for equipment, men, or funds for the purpose of developing antisubmarine warfare capabilities against Soviet ballistic missile submarines—apparently another unilateral gesture designed to assure the Soviets of a second-strike capability against the United States, under the policy of assured destruction." Patrick Glynn, *Closing Pandora's Box: Arms Races, Arms Control, and the History of the Cold War* (New York, NY: New Republic/Basic Books, 1992), 203.

"[Soviet] Marshal Zhukov, writing in Red Star in early 1957. . . said that carriers were only useful for first-strike missions since they were so vulnerable. Accordingly, they were only of interest to aggressor states like the United States and Great Britain. This has become the standard Soviet line in a persistent effort to discredit aircraft carriers." Robert Waring Herrick, *Soviet Naval Strategy: Fifty Years of Theory and Practice* (Annapolis, MD: United States Naval Institute, 1967), 70–71.

"As the decade [of the 1960s] closed the Navy's mood was effectively summarized by Vice Admiral Ruthven Libby's lament: 'We'll never recover from McNamara.' " Hegmann, "Reconsidering," 317.

"The first priority mission of naval operations in the oceanic and sea theaters will be the destruction of atomic missile submarines." V. Sokolovskiy and M. Cherednichenko, "Military Art at a New Stage," *Krasnaya Zvezda*, August 28, 1964.

"On 30 June 1970, Adm. Zumwalt reports that he had between 45 and 55 percent chance of winning a conventional war with the Soviet Union. In the spring of 1973, however, Adm. Zumwalt estimated the probability of winning against the Soviet Navy as only 25 percent if programmed end-fiscal year (FY) 1974 forces were maintained.' " In Jeffrey I. Sands, *On His Watch: Admiral Zumwalt's Efforts to Institutionalize Strategic Change*, CRM 93–22 (Alexandria, VA: Center for Naval Analyses, July 1993), 12–13.

"In 1982, then-retiring Admiral Hyman Rickover was asked how long U.S. carriers could survive in an all-out war. he said, 'About two days.' " "Are big Warships Doomed?" *Newsweek,* May 17, 1982, 35.

"Mr. Lehman's strategy . . . is costly in every sense . . . it's dangerous . . . rationales belie common sense. . . . Russia would then have few options except the first use of nuclear weapons . . . hair trigger elements of the arms race...strategy of imperial naval power . . . not clear America can afford it. . . ." James A. Nathan, "Leaky Naval Strategy," *New York Times*, January 26, 1983, 23.

"The White Paper [*Forward from the Sea*] did not abandon open ocean warfare or protection of sea lines of communication, it assumed them." Bradd Hayes, "Keeping the Naval Services Relevant," United States Naval Institute *Proceedings,* (October 1993): 59.

"[T]he peculiar psychology of the Navy Department, which frequently seemed to retire from the realm of logic into a dim religious world in which Neptune was God, Mahan his prophet, and the United States Navy the only true Church." Henry Stimson, quoted in *Time*, February 9, 1948. http://www.time.com/time/magazine/article/0,9171,855981,00.html?iid=chix-sphere (August 21, 2008).

Chapter 8: Retrospective

"No American army in 1944 would have fought the Germans in France without permission to cross the Rhine or to bomb Berlin at will. Japan would have won World War II had the United States simply fought in the jungles and occupied towns of the Japanese empire, promising not to bomb Tokyo, mine its harbors, attack its sanctuaries, or invade its native possessions, while journalists and critics visited Tokyo and broadcast to American troops from Japanese radio stations. Neither Truman nor Roosevelt would have offered to negotiate with Hitler or Stalin after the successful Normandy landings or the devastating bombing campaign over Tokyo in March 1945. GIs in World War II were killed in pursuit of victory, not in order to avoid defeat or to pressure totalitarian governments to discuss an armistice. In war it is insane not to employ the full extent of one's military power or to guarantee to the enemy that there are sanctuaries for retreat, targets that are off limits, and a willingness to cease operations anytime even the pretext of negotiations is offered." Hanson, *Carnage and Culture*, 430.

"From the earliest days of its founding, the nation has been guided by a philosophy that social historians call the American Creed. The creed's paramount values are self-reliance, stoicism, courage in the face of adversity, and the valorization of excellence." Christina Hoff Sommers and Sally Satel, *One Nation Under Therapy* (New York, NY: St. Martin's Press, 2005), 218.

"Homeland defense has seldom attracted the Navy, and seldom been required to. The Navy's posture since its earliest days has normally been forward, or trying to get forward. This has in part been to an often low threat level at or near home; a normal Navy strategy of countering threats forward; and the existence of numerous other U.S. armed forces—especially the Army and Coast Guard—with homeland defense interests and responsibilities." Peter M. Swartz, *Sea Changes: Transforming U.S. Navy Deployment Strategy: 1775–2002* (Arlington, VA: Center for Naval Analyses, 2002).

"Military opinion is naturally rigid. Men held in the grip of discipline, moving perilously from fact to fact and staking their lives at every step, are nearly always opposed to new ideas." Winston S. Churchill, *Marlborough, His Life and Times* (New York, NY: Scribner's, 1938), 107.

"Because the future has not happened, our expectations of it can only be guesswork. Historically guided guesswork should perform better than one that knows no yesterdays. Nonetheless, planning for the future, like deciding to fight, is always a gamble." Colin S. Gray, "Why Strategy Is Difficult," *Joint Force Quarterly*: (Summer 1999): 12.

"The Department of the Navy is the most strategically independent of the services—it has its own army, navy and air force. It is least dependent on others. It would prefer to be given a mission, retain complete control over all the assets, and be left alone." Gen. David Jones, quoted in James Lacy, *Within Bounds: The Navy in Postwar American Security Policy* (Arlington, VA: Center for Naval Analyses, 1983), 536.

"The soldier, few men realize, is the only one of the military men who cannot do his part of the war alone. The airman . . . does not need the soldier or the sailor to help him. The sailor can sail away and sink the enemy ships and control the seas and even extend his influence ashore, all with his own ships and his built-in air strength and his own specialized troops in the naval service. But the soldier cannot function alone. His flanks are bare, his rear is vulnerable, and he looks aloft with a cautious eye. He needs the airman and the soldier for his own security in doing his own job." Wylie, *Military Strategy*. 46.

"The army has long understood that to be successful in battle, its ground forces must be supported by other branches and services, and cannot even reach battlefields overseas without the aid of the other two services. Thus, historically its strategy has been based on an integrative, joint approach. But where the air force bases its claims to resources on advancing technological development, the army tends to emphasize the human dimension of war, and lobbies for resources to meet the needs of the nation's soldiers and their families. Different from both of the above, the navy emphasizes tradition and independence, as befits a service whose forces are 'over the horizon' much of the time and whose personnel remain focused on 'going to sea.' Hence, the navy's strategic culture has long emphasized America's insularity and reliance on overseas trade, and based its claim to resources on the need to maintain control over (and under) vast oceanic expanses. Thus, 'military' culture and 'naval' culture have been, of necessity, quite different. Manifestations of these different beliefs and attitudes have been repeatedly documented, most recently in the high-tech aspects of command and control of joint forces." Snider, "An Uninformed Debate."

"Slowly, the senior Army and Air. Force officers in this theater are realizing that there is more to naval power than is apparent to the casual yachtsman, that the sea is a hard mistress to be treated with diffidence and respect. These officers seem to fear the sea, want no part of it, and yet deep down in their hearts think they would like to command seaborne operations. We cannot hope for an overnight awakening. The principle of centralized control, issuing orders to field commanders with no prior consultation, is almost sacred. . . . [T]here had to be a continual representation, all on a very friendly basis, to convince an Army staff that naval warfare is different from land warfare. This might have turned out to have been extremely serious had naval forces been directly under the control of an Army staff. If the Navy wishes to survive, it must beware of the kindly efforts of our friendly sister services to force us to operate in the manner which is efficient for an Army—but would be deadly for a Navy." Arleigh Burke, "Burke Speaks Out on Korea," Naval Institute *Proceedings* (May 2000): 68–69.

"'We can't do things unilaterally, we recognize that,' says Donald Winter, the Navy Secretary. 'Not all things, not all places.' " Gordon Lubold, "U.S. Navy Aims To Flex 'Soft Power,' " *Christian Science Monitor*, December 27, 2007, 2.

"Just after the Spanish-American War, Alfred Thayer Mahan threw the chasm between service mindsets into bold relief when he wrote: 'If we lost ten thousand men, the country could replace them; if we lost a battleship, it could not be replaced.' Three generations later, Admiral James Watkins said, 'I like interservice rivalries. In peacetime I could just never, never bring myself to do anything to help the Air Force or Army." Quoted in Roger A. Beaumont, *Joint Military Operations: A Short History* (Westport, CT: Greenwood Press, 1993), 190.

"These differences of judgment, these clashes of ideas, these almost constant pullings and haulings among the services, are the greatest source of military strength that the nation has. We do differ, within and among the services, and may Heaven help us if we ever enter into a period of sweetness and light and unanimity." J. C. Wylie, "Why a Sailor Thinks Like a Sailor," U.S. Naval Institute *Proceedings* (August 1957): 812.

A destroyer, even the brave might fear,
She inspires horror in the harbor and the open sea,
She goes into the waves flanked by arrogance, haughtiness, and fake might,
To her doom she progresses slowly, clothed in a huge illusion,
Awaiting her is a dinghy, bobbing in the waves.

Poem by Osama bin Laden after the bombing of the *Cole*. Quoted in Lawrence Wright, *The Looming Tower: Al Qaeda and the Road to 9/11* (New York, NY: Vantage Books, A Division of Random House, Inc., 2006), 376.

"At Wonsan harbor naval mines delayed a robust U.S./NATO amphibious assault force of 50,000 marines supported by 250 ships, for twenty days. Reacting to this successful denial of access by the North Koreans, CATF [Commander of the Amphibious Task Force] declared to the CNO [Chief of Naval Operations]: 'We have lost control of the seas to a nation without a Navy, using pre-World War I weapons, laid by vessels that were utilized at the time of the birth of Christ.' " Charles E. Wilhelm, "Forward . . . From the Sea: The Mine Warfare Implications," *Marine Corps Gazette* 79 (July 1995): 23.

"This illustrates the trouble with using mines as antiaccess weapons; in most cases, those who seek to deny access do not have the means to lay enough mines to make a major difference." William D. O'Neil, "The Naval Services: Network-Centric Warfare," in Binnendijk, *Transforming*, 140.

"When he was asked what the lesson of the Gulf War was, the Indian chief of staff is reported to have said, 'Never fight the United States without nuclear weapons.' " Philip Bobbitt, "Law, Strategy, and History," in Richard Little and Michael Smith, Eds, *Perspectives on World Politics*, 3d ed. (London: Routledge, 2006): 116.

"One of a carrier's virtues is its potential to be employed early in a period of crisis. Given the new strategic environment (i.e., the absence of a significant blue-water challenger) this will require the carrier to move ever closer to shore . . . as the carrier's shorter-range aircraft require. Yet this means putting the lives of the carrier's 5,000-6,000 person crew at ever greater risk, for what may amount to interests far less than vital to U.S. survival or well-being." Andrew Krepinevich, *A New Navy for a New Era*. Center for Strategic and Budget Assessments, 1996. http://www.csbaonline.org/4Publications/PubLibrary/R.19960500.A_New_Navy_For_A_N/R.19960500.A_New_Navy_For_A_N.php. (August 22, 2008).

"If the Navy is to make good on its avowed mission to be first on the scene in the event of crisis, and to provide a prompt global strike capability when the nation requires it, then it will have to find some way of doing it without putting so many of its sailors in harm's way, especially for those contingencies where the stakes are relatively low." Andrew Krepinevich, *A New Navy for a New Era*. Center for Strategic and Budget Assessments, 1996. http://www.csbaonline.org/4Publications/PubLibrary/R.19960500.A_New_Navy_For_A_N/R.19960500.A_New_Navy_For_A_N.php (August 22, 2008).

"Anything that can be located in time and space can be targeted and destroyed, and the only limitations on locating the target are the expense, effort, and time the attacker can accept in solving the problem." Mark C. Lewonowski, "Information War," in Thomas C. Gill, ed., *Essays on Strategy IX* (Washington, D.C.: National Defense University Press, 1993), 59.

"I would suggest, therefore, that maritime units have as much, perhaps more, flexibility than any other platform in managing operational risk. Of course some ships may suffer hits, but during the Falklands conflict a large number of RN escorts demonstrated that it is possible for a surface ship to absorb considerable punishment, survive, recover, and then continue contributing to the task. In this, a ship compares very favourably with platforms such as aircraft or armoured vehicles. My point is that ships are designed, procured, equipped and trained to be exposed to operational risk and to manage the consequences. Indeed, they are among the most capable all-round risk managers in a nation's defence inventory." Admiral Sir Benjamin Bathurst, Chief of Naval Staff and First Sea Lord, "Naval Aspects of the Post-Cold War Era," *Defence Systems International,* 1994–1995, 109.

"The risk to carriers that steam into harm's way is rising, and will likely continue to do so. Offsetting the risk by looking for a 'carrier-centered' solution will likely prove prohibitively expensive." Andrew Krepinevich, *A New Navy for a New Era*. Center for Strategic and Budget Assessments, 1996. http://www.csbaonline.org/4Publications/PubLibrary/R.19960500.A_New_Navy_For_A_N/R.19960500.A_New_Navy_For_A_N.php (August 22, 2008).

"Clausewitz, like the Mahans, believed that command had to be exercised in the face of uncertainty and thus required moral as well as intellectual qualities." Sumida, *Inventing*, 112.

"Naval forces have been intervening in land wars time out of mind. By the 17th century, nations had begun to invest heavily in coastal defenses to prevent this. Fortifications, .seacoast artillery, and physical barriers were built. Ever since then, the impossibility of breaching seacoast defenses has repeatedly been asserted and repeatedly been proven wrong." William D. O'Neil, "The Naval Services:

Network-Centric Warfare," in Hans Binnendiyk, ed., *Transforming America's Military* (Washington, D.C.: National Defense University Press, 2002), 135.

A QUICK MOVEMENT BY THE ENEMY WILL JEOPARDIZE SIX FINE GUNBOATS (A Pangram)

"Only fools and passengers drink at sea." Allan Villiers. http://www.armstronglobal.com/quotes.html (August 22, 2008).

"In short, for reason to tolerate those who refuse to play by the rules of reason is nothing else but the suicide of reason—and with the suicide of reason, mankind will face the dismal prospect of a return to the brutal law of the jungle that has governed human communities for the vast bulk of both our history and our prehistory, and from which certain lucky cultures have miraculously managed to escape—and even then, only by the skin of their teeth." Lee Harris, *The Suicide of Reason: Radical Islam's Threat to the West* (New York, NY: Basic Books, 2007), 278–279.

"Lionel Tiger . . . argued that men tend to bond most strongly in situations involving power, force, and dangerous work, and they consciously and emotionally exclude females from these groups." Browne, *Co-Ed Combat*, 161–162.

"Manliness is knowing how to be confident in situations where sufficient knowledge is not available. . . . Manliness is a passionate quality, and it often leads to getting carried away, whether for good or ill. A sober, scholarly treatment risks failing to convey the nobility of manliness. . . . A manliness that seeks glory in risk and cannot abide the rational life of peace and security." Mansfield, *Manliness,* 21.

Love's fiery beams I cannot smother,
A kiss of a seaman's worth two of another.
"The Kiss of a Seaman," Anonymous, Roxburghe Ballads. http://www.julian-stockwin.com/Poetry.htm (August 22, 2008).

" 'In berthing the guys mainly read porno,' she said. 'There are too many fat girls, too many dykes, there's no possibility of a bath, I give myself a pedicure once a week to remember what I am. I got punished for kissing another deck ape in Hawaii—he was just a good friend.' " Robert D. Kaplan, *Hog Pilots, Blue Water*

Grunts: The American Military in the Air, at Sea, and on the Ground (New York, NY: Random House, 2007), 123.

"That doughty sixteenth-century French warrior Blaise de Montluc had warned that the love of women was 'utterly an enemy to an heroic spirit,' and a long series of martial groups have created and preserved all-male environments in which sex has—at least in theory—been sublimated or repressed." Holmes, *Acts of War*, 94.

"A woman needs to be told that you would sacrifice anything for her. A man needs to be told he is being useful. When the man or woman strays from that formula, the other loses trust. When trust is lost, communication falls apart." Scott Adams, *God's Debris: A Thought Experiment* (Kansas City, KS: Andrews McMeel Publishing, 2004), 112.

"In a small and isolated group with a skewed sex ratio, sex is a resource that cannot be shared equally and therefore almost always harms cohesion." Browne, *Co-Ed Combat*, 195.

"Men and women interpret aggression differently: Women see it as a loss of self-control and are ashamed of their anger, associating it with being pushy, nasty, and socially isolated. Men, by contrast, see their aggressiveness in a positive light, as a way of gaining control." David P. Barash, "Evolution, Males, and Violence," *The Chronicle of Higher Education* (May 24, 2002). http://chronicle.com/free/v48/i37/37b00701.htm (April 24, 2008).

"Male bonding is explicitly reinforced by Nelsonian imagery, particularly by the notion of the 'band of brothers' (in which close that all involved) instinctively understand how colleagues close that all involved instinctively understand how colleagues will react under the testing circumstances of war. The Royal Navy actively encourages officers from the start of their careers to not only work together but also to be involved in sport and socialize or in other words to spend as much time together as possible. Consequently the levels of interaction are far more intimate than in most other professions, more akin to family than work-mates, and all aspects of personal as well as professional life are known to all. The issue of gender, however, has injected new problems within an institution that promotes close associations between colleagues as the recipe for success." Alastair Finlan, *The Royal Navy in the Falklands Conflict and the Gulf War: Culture and Strategy* (London: Frank Cass, 2004), 13.

"The gender-neutral society is not friendly toward risky activity, even on behalf of liberty, that might give advantage to manly men as risk-takers and thus upset the balance of the sexes on which it depends." Mansfield, *Manliness*, 166.

"The fact is that people learn what you teach them. And the consequences of the war against boys—and the broader social war against masculinity in general—are increasingly evident in both the culture and the world at large. We should hardly be surprised that the results are anything but pretty." S. T. Karnick, "Girly Men: The Media's Attack on Masculinity," *Salvo Magazine* 4 (2008).

"We have taught a generation of soldiers to see themselves not primarily as soldiers, but as African-Americans who happen to be soldiers, or females who happen to be soldiers. Worse yet, we have taught them not to be polite and respectful, but instead to carry chips on their shoulders, searching for someone to offend them. The result in the loss of unit cohesion has been devastating as soldiers are isolated in social fear." Letter to Center for Strategic and International Studies on its Study: "American Military Culture in the 21st Century," from Jim Carson, Aviation Regiment CMR 477 Box 1551 APO AE 09165, 27 March 2001. http://royator.org/ator/Military_Culture.html (August 22, 2008).

"America has, and has had, a very successful military, and to debate policies designed to change its culture without at the same time having an informed discussion of the consequences in terms of military effectiveness is folly." Snider, "Uninformed."

Military society "is characterized by its own laws, rules, customs, and traditions, including numerous restrictions on personal behavior that would not be acceptable in civilian society." *Military Personnel Eligibility Act of 1993*, 10 USC § 654.

"All-male groups have existed in virtually every known society. Most anthropologists agree that all-male groups produce a peculiar kind of nonerotic psychological bond that men crave and cannot find elsewhere." Mitchell, *Women in the Military*, 174.

"Somewhere in that soldier's world view, though he may not be able to articulate it, is the notion that he is here . . . so that all the higher ideals of home embodied in mother, sister, and girlfriend do not have to be here." Neil L. Golightly, "No Right to Fight," U.S. Naval Institute *Proceedings* (December 1987), 48.

"The military can attempt to drive out the elements of human nature that made integration of combat units so difficult, but in the end, nature will exact a heavy toll." Browne, *Co-Ed Combat*, 287.

"Strip the male of his protective role—rooted in his nature and codified throughout the centuries—and you just can't predict the consequences. . . . Wherein lies potential tragedy. . . . That very modern game—pretending we're all the same—costs us all in the end, being built on a damnable lie." William Murchison, "Sex, lies . . . and the U.S. Navy *The Washington Times*, August 4, 1992, F3.

"Women should also be barred from warships. Problems of pregnancy and sexual relations and the impact on cohesion that they can have pose a substantial threat to military effectiveness. In war and peace, but especially in war, all hands must be prepared to pitch in and perform the highly physical task of damage control. Women's lack of upper-body strength puts the entire ship's crew at risk. Although we have not faced a serious naval power in combat since World War II, we may in the future, and the problems will become even more serious." Browne, *Co-Ed Combat*, 297.

"Warfare is a supranational survival contest in which opposing sides vie for any advantage; unilateral policies adopted to promote principles other than military necessity may place the adopting party at increased risk of defeat." Elaine Donnelly, "Constructing the Co-Ed Military," *Duke Journal of Gender Law and Policy* 14:815, 899.

"In any society where enlightened self-interest rules, the heroism that is willing to face certain death becomes a moral anachronism." Harris, *Suicide of Reason*, 114.

"It will avail us little if the members of our defeated force are all equal. History will treat us for what we were: a social curiosity that failed." Richard A. Gabriel, quoted in Mitchell, *Women in the Military*, 333.

"Only the guy who isn't rowing has time to rock the boat." Jean-Paul Sartre. http://www.cognitivedistortion.com/?cd=quotes&aid=8169 (August 22, 2008).

Ye gentlemen of England
That live at home at ease
Ah! little do you think upon
The dangers of the seas
Song by Martyn Parker, quoted in Heinl, *Dictionary*, 206.

Chapter 9: Conclusion

They that go down to the sea in ships,
that do business in great waters;
these see the works of the LORD,
and his wonders in the deep.
Psalm 10.
"The sea drives truth into a man like salt." Hillarie Belloc. http://www.sailnet.com/forums/mass-bay-sailors/34849-sea-drives-truth-into-man-like-salt.html (August 17, 2008) (August 23, 2008).

"The Navy is the asylum for the perverse, the home of the unfortunate. Here the sons of adversity meet the children of calamity, and here the children of calamity meet the offspring of sin." Herman Melville, *White Jacket or the World in a Man-of-War* (New York, NY: Viking Press, 1983), 425.

"Sailors, with their built-in sense of order, service, and discipline, should really be running the world." Nicholas Monsarrat. http://www.indigorising.net/phpBB2/viewtopic.php?t=2632&sid=f59adc0d59adfe8113570cc8cc0d1b9e (August 23, 2008).

"The vitality and morale of one's culture (which include a sense of its future possibilities) are the keys to civilizational success and influence over the long haul." George Weigel, *Faith, Reason, and the War against Jihadism: A Call to Action* (New York, NY: Doubleday, 2007), [page number omitted].

"This concern with economic interests throughout the world was not something that developed when America passed, in Mr. McGeorge Bundy's phrase, 'from Innocence to Engagement.' America was always 'engaged.' The offer of naval forces abroad was as old as the U.S. Navy itself. . . . America's emergence as a world power was a steady, dynamic affair; the warship only policed the process." Karsten, *Naval Aristocracy,* 172.

"A fundamental enabler of globalization in most regions of the world is the absence of a major war or major internal strife. A stable, secure environment is often taken for granted, but it is the underpinning of growth. Since sound business decisions require a degree of stability, investments tend to be postponed when nations are at war or on the brink of war." Ellen Frost, "Globalization and National Security: A Strategic Agenda," in Richard L. Kugler and Ellen L. Frost, eds., *The Global*

Century: Globalization and National Security, (Washington, D.C.: National Defense University Press, 2001), 42.

"Weakness provokes insult and injury, while a condition to punish often prevents them. This reasoning leads to the necessity of some naval force; that being the only weapon by which we can reach an enemy." Thomas Jefferson, *The Jefferson Cyclopedia*, no. 5784 (New York, NY: Funk and Wagnalls Company, 1900), 619.

"Globalization and the strategic thinking articulated by Alfred Thayer Mahan go hand in hand. Mahan's vision of a United States growing rich from its ability to use the seas as a means of communication fits well with contemporary thinking about how the information revolution has facilitated international commerce, contacts among individuals, and cultural exchange. The Navy plays a critical role in the process of globalization because it controls access to the world's primary means of communication (ocean transportation) and, by implication, access to global resources and markets. The Navy guarantees that the United States, its allies, and its friends will have access to the wealth produced by global trade among market economies. The Navy helps to create and maintain the political, commercial, and security conditions necessary for globalization to occur. The Navy patrols and protects the sea lines of communication/commerce that spread democracy and create global markets." James J. Wirtz, "Will Globalization Sink the Navy?" in Sam Tangredi, ed., *Globalization and Maritime Power* (Washington, D.C.: National Defense University Press, 2002). http://www.ndu.edu/inss/books/Books_2002/Globalization_and_Maritime_Power_Dec_02/31_ch30.htm (August 23, 2008).

"Although the sailor is no less, and one can hope no more, partisan than any other military man, no sailor is so naïve as to suppose that the Navy alone is going to sail out and win all our wars." Wylie, "Why a Sailor Thinks Like a Sailor," 817.

"The history of failure in war can be summed up in two words—too late. Too late in comprehending the deadly purpose of a potential enemy; too late in realizing the mortal danger; too late in preparedness; too late in uniting all possible forces for resistance; too late in standing with one's friends." Douglas MacArthur, 1941, quoted in Senator Joseph Lieberman, "Winning the Wider War against Terrorism," Lecture delivered at Georgetown University, January 14, 2002. http://www.yale.edu/lawweb/avalon/sept_11/lieberman_001.htm (August 23, 2008).

"Want of foresight, unwillingness to act when action would be simple and effective, lack of clear thinking, confusion of counsel until the emergency comes, until self-preservation strikes its jarring gong—these are the features which constitute the endless repetition of history." Winston Churchill, To the House of Commons, May 2, 1935. http://quotationsbook.com/quote/add_to_site/45524/ (August 23, 2008).

"[T]oday, in the post 9/11 world, we know now that catastrophic terror is a possibility, and this knowledge can never be eluded. It changes the way in which we imagine our future, just as it changed the way those who had lived through the Great War were condemned to imagine theirs." Harris, *Civilization and its Enemies*, 55.

"Any step that is not good for the Navy is not good for the nation." Fleet Admiral E. J. King, Senate Hearings 23 Oct 1945. Quoted in Paolo E. Coletta, *The United States Navy and Defense Unification, 1947–1953* (Newark, DE: University of Delaware Press, 1981), 177.

"Freedom of the seas in peace and sea supremacy in war are basic to the existence of the United States as a world power. They are keystones of United States foreign policy and are essential to the Free World cooperation which it nurtures." Arleigh Burke, "Origins of United States Navy Doctrine," Attachment to Letter to All U.S. Naval Officers, Officer Candidates, Midshipmen and Cadets (April 11, 1960), 11.

"The American people, he [Mahan] declared in 1912, 'are singularly oblivious of the close relation between peace and preparation.' " Samuel Huntington, *The Soldier and the State* (Cambridge, MA: Harvard University Press, 1957), 279.

"They do not realize that the Army is so absolutely different from the Navy. Every condition in them both is different. The Navy is always at war, because it is always fighting winds and waves and fog. The Navy is ready for an instant blow. . . . The ocean is limitless and unobstructed; and the fleet, each ship manned, gunned, and provisioned and fuelled, ready to fight within five minutes." Sir John Fisher, *Memories*, 1919. Quoted in Heinl, *Dictionary*, 210.

"Strategically, then, as well as politically, navies and their admirals tend to favor the status quo." Clark G. Reynolds, *History and the Sea: Essays on Maritime Strategies* (Columbia, SC: University of South Carolina Press, 1989), 8

"It is dangerous to meddle with Admirals when they say they can't do things." Winston Churchill, quoted in Edward J. Marolda, *FDR and the U.S. Navy* (New York, NY: Macmillan, 1998), 165.

"The Soviet Union's most deadly forces, its ICBMs and tank armies, were relatively easy to find, but hard to kill. Intelligence was important, but overshadowed by the need for sheer firepower. Today, the situation is reversed. We are now in an age in which our primary adversary is easy to kill, but hard to find. You can understand why so much emphasis in the last five years has been on intelligence." *Remarks of Central Intelligence Agency Director Gen. Michael V. Hayden* at the Council on Foreign Relations (as prepared for delivery), September 7, 2007.

"Many of my friends are now dead. To a man, each died with a nonchalance that each would have denied as courage. They simply called it lack of fear. If anything great or good is born of this war, it should not be valued in the colonies we may win nor in the pages historians will attempt to write, but rather in the youth of our country, who never trained for war; rather almost never believed in war, but who have, from some hidden source, brought forth a gallantry which is homespun, it is so real. When you hear others saying harsh things about American youth, do all in your power to help others keep faith with those few who gave so much. Tell them that out here, between a spaceless sea and sky, American youth has found itself and given itself so that, at home, the spark may catch. There is much I cannot say, which should be said before it is too late. It is my fear that national inertia will cancel the gains won at such a price. My luck can't last much longer, but the flame goes on and on." Ensign William R. Evans, USN, a pilot of Torpedo Squadron 8, KIA at Midway, 4 June 1942. The Battle of Midway Roundtable. http://home.comcast.net/~r2russ/midway/ (August 23, 2008)

"Men go into the navy . . . thinking they will enjoy it. They do enjoy it for about a year, at least the stupid ones do, riding back and forth quite dully on ships. The bright ones find that they don't like it in half a year, but there's always the thought of that pension if only they stay in. So they stay . . . gradually they become crazy. Crazier and crazier. Only the Navy has no way of distinguishing between the sane and the insane. Only about 5 percent of the Royal Navy have the sea in their veins. They are the ones who become captains. Thereafter, they are segregated on their bridges. If they are not mad before this, they go mad then. And the maddest of these become admirals." Attributed to George Bernard Shaw, quoted in Heinl, *Dictionary,* 210.

"Nothing in the world, nothing that you may think of, or dream of, or anyone may tell you; no arguments, however specious; no appeals, however seductive, must lead you to abandon that naval supremacy on which the life of our country depends." Winston S. Churchill, quoted in Benjamin H. Williams, *The United States and Disarmament* (New York, NY: McGraw-Hill Book Company, Inc., 1931), 137.

"The cure for everything is salt water—sweat, tears or the sea." Isak Dinesen. http://saltwaterchronicles.blogspot.com/ (August 23, 2008).

"For you [the United States], the military is not a question of life and death. . . . So you can afford to make all kinds of social experiments, which we [Israel] cannot. . . . The very fact that you have this debate may itself be construed as proof that it's not serious. It's a game. It's a joke." Martin Van Creveld, quoted in Mitchell, *Women in the Military*, 215

"Although the military defends the principles of democratic society, it cannot fully embody them. Its end is victory, not equity; its virtue is courage, not justice; its structure is authoritarian, not pluralistic. . . . The requirements of military life clash with the democratic commitment to equality, natural rights, and consent." Jean Yarbrough, "The Feminist Mistake: Sexual Equality and the Decline of the American Military," *Policy Review* (Summer 1985), 52.

"A Navy is organized to gain victory at sea, not to illustrate ideas about human equality." C. Northcote Parkinson, "Comments on 'democratization' of Royal Navy Officer Procurement," in *New York Times*, August 20, 1961.

"The nation that will insist on drawing a broad line of demarcation between the fighting man and the thinking man is liable to find its fighting done by fools and its thinking done by cowards." Sir William Francis Butler. Courtesy of *Eigen's Political & Historical Quotations*. http://www.politicalquotes.org/Quotedisplay.aspx?DocID=59519 (August 23, 2008).

"The dependence of the American people upon the sea—to utilize it for political and economic power and even for cultural growth—has been so dramatically demonstrated throughout our history as to make the image of the republic incomprehensible without it. Without American activity upon the waters, alone and with allies, how different the course of American and world history would have been—from the crossing of the Pilgrim fathers on the Mayflower to the recovery of the Apollo astronauts." Clark G. Reynolds, "The Sea in the Making of America," *To Use The Seas: Readings in Seapower and Maritime Affairs* (Annapolis: Naval Institute Press, c. 1977), 31.

Notes

Preface

1. Quoted in Bradley Peniston, *No Higher Honor: Saving the USS* Samuel B. Roberts *in the Persian Gulf* (Annapolis, MD: Naval Institute Press, 2006), 190.
2. James Holmes, "Why Doesn't America Have a Nelson?" *Naval War College Review* 58, no. 4 (Autumn 2005), 19–20.
3. Don M. Snider, "An Uninformed Debate on Military Culture," *Orbis* 43, no. 1 (Winter 1999).
4. Samuel P. Huntington, *The Soldier and the State: The Theory and Politics of Civil-Military Relations* (Cambridge, MA: Harvard University Press, 1957), 74.
5. Harvey Mansfield, "Man of Courage, Alexander Solzhenitsyn, 1918–2008," *Weekly Standard* 013, no. 46 (August 25, 2008).

Chapter 1. Introduction

1. Clark G. Reynolds, *Command of the Sea: The History and Strategy of Maritime Empires* (Malabar, FL: Robert E. Krieger Publishing Company, 1983), 1:66.
2. Edwin Hoyt, quoted in Samuel P. Huntington, "Playing to Win," *The National Interest* (Spring 1986): 8.

3. Jules Jusserand, quoted in Thomas A. Bailey, *A Diplomatic History of the American People*, 6th ed. (New York, NY: Appleton-Century-Crofts, Inc., 1958), 4.
4. Reynolds, *Command of the Sea*, 5.
5. A theme investigated at length in Colin S. Gray, *The Leverage of Sea Power: The Strategic Advantage of Navies in War* (New York, NY: The Free Press, 1992).
6. A "high-ranking Air Force officer," quoted in Samuel P. Huntington, "National Policy and the Transoceanic Navy 80, no. 5 (May 1954), 484.
7. Air Force Gen. H. J. Knerr, quoted in Steven E. Miller, "Rough Sailing: The U.S. Navy in the Nuclear Age," in *America's Defense,* ed. Michael Mandelbaum (New York and London: Holmes and Meier, 1989), 198.

Chapter 2. Strategic Culture

1. A point addressed cogently years ago by Huntington, *Soldier and the State*, 61–62.
2. See, for example, Peter M. Swartz, *"Forward . . . From the Start": The U.S. Navy & Homeland Defense: 1775–2003.* COP D0006678.A1/Final. (Arlington, VA: Center for Naval Analyses, February 2003).
3. In W. Spencer Johnson, "New Challenges for the Unified Command Plan," *Joint Force Quarterly* (Summer 2002), 63.
4. This was inspired by Lee Harris, *Civilization and Its Enemies*: *The Next Stage of History* (New York, NY: The Free Press, 2004), 106.
5. DA Form 71, 1 August 1959, for officers.
6. SECNAV Instruction 1000.9.
7. The code can be found at: http://www.au.af.mil/au/awc/awcgate/ucmj.htm. (August 16, 2008).
8. See Roger W. Barnett, *Asymmetrical Warfare* (Washington, DC: Brassey's, 2003). This book describes the myriad of constraints levied on the ability of U.S. military forces to employ military power. It suggests that while any given constraint might be proper and even desirable, their cumulative weight tilts the battlespace in favor of adversaries who operate entirely free of such constraints.
9. Harris, *Civilization and Its Enemies*, 107.
10. For an in-depth, well-balanced elucidation, see Snider, "Uninformed Debate."

11. Donald M. Schurman, "Mahan Revisited," in *Maritime Strategy and the Balance of Power,* eds. John Hattendorf and Richard Jordan (New York, NY: St. Martin's Press, 1989), 106.
12. Lt. Gen. Sir Hew Pike, quoted in Ernest Blazar, "Glass Houses," *Washington Times*, December 1, 1997, 8.
13. http://s100megsfree4.com/napwars/maxims12.html (August 16, 2008).
14. Samuel Eliot Morison, "Notes on Writing Naval (not Navy) English," *American Neptune* (1949) 9:10.
15. Henry A. Leonard, Michael Polich, Jeffrey D. Peterson, Ronald E. Sortor, and S. Craig Moore, *Something Old, Something New: Army Leader Development in a Dynamic Environment,* MG-280 (Santa Monica, CA: RAND, 2006), xv.
16. Julian S. Corbett, *Some Principles of Maritime Strategy (*Annapolis, MD: Naval Institute Press, 1988), 167.
17. Barry M. Gough, "The Influence of History on Mahan," in *The Influence of History on Mahan,* ed. John B. Hattendorf (Newport, RI: Naval War College Press, 1991), 22.
18. Peter Morville, *Ambient Findability* (Sebastopol, CA: O'Reilly, 2005), 169.
19. Richard Holmes, *Acts of War: The Behavior of Men in Battle* (New York, NY: The Free Press, 1985), 8.
20. Alan Charles Kors, "The West at the Dawn of the 21st Century: Triumph Without Self-Belief," *Watch on the West* 2, no. 1 (Philadelphia, PA: Foreign Policy Research Institute, February 2001), 4.
21. Thucydides, *The History of the Peloponnesian War,* http://classics.mit.edu//Thucydides/pelopwar.html (August 15, 2008).

Chapter 3. The Maritime Context

1. Donald. A. Merlin, *A Mind So Rare: The Evolution of Human Consciousness* (New York, NY: W. W. Norton & Company, 2001).
2. John O. Coote, ed., *The Norton Book of the Sea* (New York, NY: W. W. Norton & Company, 1989), 4.
3. Victor Davis Hanson, *Carnage and Culture: Landmark Battles in the Rise of Western Power* (New York, NY: Doubleday, 2001), 28.
4. Dava Sobel, *Longitude: The True Story of a Lone Genius Who Solved the Greatest Scientific Problem of his Time* (New York, NY: Walker, 1995).
5. John Masefield, *The Bird of Dawning; Or, the Fortune of the Sea* (New York, NY: MacMillan Company, 1933), 34.

6. J. H. Parry, *The Discovery of the Sea* (Berkeley, CA: University of California Press, 1981; first pub. 1974), xi.
7. A notion expanded on by Colin S. Gray in his *Recognizing and Understanding Revolutionary Change in Warfare: The Sovereignty of Context* (Carlisle, PA: U.S. Army War College Strategic Studies Institute, February 2006).
8. *The Commander's Handbook on the Law of Naval Operations*, NWP 1-14M, (Washington, DC: Office of the Chief of Naval Operations, 2007), Section 4.7.1.
9. Bernard H. Oxman, "The Territorial Temptation: A Siren Song at Sea," *American Journal of International Law* 100:830), 841.
10. President Ronald Reagan, "Excerpts from the Text of Remarks by the President at the Recommissioning of the USS *New Jersey*, Long Beach CA, December 28, 1982," (The White House, Office of the Press Secretary, December 29).
11. A. T. Mahan, *The Problem of Asia and Its Effect upon International Policies* (Boston, MA: Little, Brown, and Company, 1905), 38.
12. See, for example, Stephen B. Olsen, "Developing a Coastal Husbandry," *Providence Journal-Bulletin* (February 29, 1996), B6.
13. Richard Owen and Daniel McGrory, "Business-Class Suspect Caught in Container," *London Times* (October 25, 2001).
14. Yaneer Bar-Yam, *Making Things Work: Solving Complex Problems in a Complex World* (NECSI Knowledge Press, 2004), 99.

Chapter 4. Strategies for the Employment of Naval Forces

1. Bernard Brodie, *A Layman's Guide to Naval Strategy* (Princeton, NJ: Princeton University Press, 1943), 12–13.
2. In particular, Mahan's *The Influence of Sea Power upon History 1660–1783*, 12th ed. (Boston, MA: Little, Brown, 1918); Philip Colomb's *Essays on Naval Defence* (London: W. H. Allen & Co., Ltd., 1893); and Julian Corbett's *Some Principles of Maritime Strategy* (Annapolis, MD: Naval Institute Press, 1988, reprint of 1911 edition).
3. B. H. Liddell Hart, 1950, quoted in Robert Debs Heinl Jr, *Dictionary of Military and Naval Quotations* (Annapolis, MD: United States Naval Institute, 1966), 208.
4. Based on Russell Spurr, *A Glorious Way to Die: The Kamikaze Mission of the Battleship Yamato, April 1945* (New York, NY: Newmarket Press, 1981).

5. A recent compilation of blockade strategies is: *Naval Blockades and Seapower: Strategies and Counter-Strategies, 1805-2005,* eds. Bruce A. Elleman and S. C. M. Paine (London and New York, NY: Routledge, 2006).
6. John Mueller and Karl Mueller, "Sanctions of Mass Destruction," *Foreign Affairs* (May/June 1999) 49–53.
7. Alfred Thayer Mahan, "Blockade In Relation To Naval Strategy," *Proceedings* 21, no. 4 (1895), 856.
8. See Roger W. Barnett, "Soviet Strategic Reserves and the Soviet Navy," in *The Soviet Union: What Lies Ahead?*, ed. Kenneth M. Currie and Gregory Varhall (Washington, DC: U.S. Government Printing Office, 1984).
9. Bernard Brodie, "Strategy as an Art and a Science," *Naval War College Review* (Winter 1998), 26–38.

Chapter 5. Expeditionary

1. "Two books addressing the issue of more European slaves being brought to North Africa than African slaves being brought to America have been published in recent years. They are *Christian Slaves, Muslim Masters* by Robert Davis and *White Gold* by Giles Milton. Both books have been largely ignored by the media and academia alike—and the first went out of print, less than 6 months after being published." Thomas Sowell, "Are Facts Obsolete?" http://townhall.com/columnists/ThomasSowell/2006/04/04/are_facts_obsolete) (August 19, 2008).
2. Quoted in Christopher Hitchens, "Jefferson Versus the Muslim Pirates," *City Journal* (Spring 2007). http://www.city-journal.org/html/17_2_urbanities-thomas_jefferson.html (August 19, 2008).
3. Joshua E. London, *Victory in Tripoli: How America's War with the Barbary Pirates Established the U.S. Navy and Built a Nation* (Hoboken, NJ: John Wiley & Sons, Inc., 2005), 23–24.
4. Osama bin Laden, "Kill Americans Everywhere," available in English translation at http://www.emergency.com/bladen98.htm (August 19, 2008).
5. Swartz, *Forward . . . From the Start.*
6. Quoted in R. O. Coulthard, "Disarmament—The World's Enigma," *Naval War College Review* (September 1960), 18.
7. Bradford Dismukes and James M. McConnell, eds., *Soviet Naval Diplomacy* (New York, NY: Pergamon, 1979).
8. According to Vice Adm. Jerry Miller, "How We Targeted the Nukes," *Proceedings* (February 2002), 44–47.

9. The statement was made after the Battle of Trafalgar, but while Napoleon was still directing his armies successfully on the Continent. Geoffrey Till, *Maritime Strategy and the Nuclear Age* (New York, NY: St. Martin's Press, 1982), 147.
10. Oxman, "Territorial Temptation," 841.
11. Wayne P. Hughes, *Fleet Tactics: Theory and Practice* (Annapolis, MD: Naval Institute Press, 1986).
12. Andre Beaufre, Strategy *of Action* (New York, NY: Frederick A. Praeger, Publishers, 1967), 121.
13. Brodie, *Layman's Guide*, 4–5.
14. Forrest Sherman, quoted in *U.S. News and World Report* (23 February 1951), 27.
15. *Sea-Based Airborne Antisubmarine Warfare 1940–1977*, vol. 1, *1940–1960.* Prepared for Op-095 under ONR contract N00014-77-C-0338 (Alexandria, VA: R. F. Cross Associates, 1978).
16. Hughes, *Fleet Tactics*, 147.
17. See, for example, Sean M. Lynn-Jones, "A Quiet Success for Arms Control: Preventing Incidents at Sea," *International Security* (Spring 1985), 154–184.
18. George Washington, letter to Bushrod, November 9, 1787. http://gwpapers.virginia.edu/documents/constitution/1787/washington.html (August 19, 2008).
19. Quoted in *The Chiefs of Naval Operations,* ed. Robert Love Jr. (Annapolis, MD: U.S. Naval Institute Press, 1980), 287.
20. Lawrence Wright, *The Looming Tower: Al Qaeda and the Road to 9/11* (New York, NY: Vantage Books, 2006), 359.
21. See Barnett, *Asymmetrical Warfare.*
22. Quoted in George Will, "The Politics of Memory," http://townhall.com/columnists/GeorgeWill/2006/08/20/the_politics_of_memory. (August 20, 2008).

Chapter 6. Technology and Systems

1. Laird G. Clowes, quoted in Geoffrey J. Marcus, *Heart of Oak: A Survey of British Seapower in the Georgian Era* (New York, NY: Oxford University Press, 1975), 10.
2. Michael E. Vlahos, "The Making of an American Naval Tradition (1795–1887)," in *Naval Engineering and American Sea Power*, ed. Millard S. Firebaugh, 2d ed. (Dubuque, IA: Kendall/Hunt Publishing Company, 2000), 3.

3. Willis C. Barnes, "The Cold War: Korea and Vietnam (1950–1972)," in Firebaugh, *Naval Engineering*, 298.
4. Reynolds, *Command of the Sea*, 6.
5. Based on Buckner F. Melton Jr. *Sea Cobra: Admiral Halsey's Task Force and the Great Pacific Typhoon* (Guilford, CT: The Lyons Press, 2007), 11.
6. http://moneycentral.communities.msn.com/HistoryWarPolitics1775/great-quotes.msnw?action=get_message&mview=0&ID_Message=288&LastModified=4675559446513873328 (August 20, 2008).
7. John Keegan, *The Price of Admiralty: The Evolution of Naval Warfare* (New York, NY: Viking, 1989), 90.
8. Two detailed essays on the subject, a fuller discussion of which need not detain this chapter, can be found in Alan Beyerchen, "Clausewitz, Nonlinearity, and the Unpredictability of War," *International Security* 17, no. 3 (Winter 1992–1993), 59–90; and Barry Watts, "Clausewitzian Friction and Future War," McNair Paper No. 52. (Washington, DC: Institute for National Strategic Studies, 1999).
9. Quoted in Peter J. Dombrowski and Andrew L. Ross, "Transforming the Navy: Punching a Feather Bed," *Naval War College Review* (Summer 2003), 107.
10. Barnes, "Cold War," in Firebaugh, *Naval Engineering*, 314.
11. Bar-Yam, *Making Things Work*, 99.
12. Vice Adm. Arthur K. Cebrowski and John J. Garstka, "Network-Centric Warfare: Its Origin and Future," *Proceedings* (January 1998), 28–35.
13. Peter Kassan, "A. I. Gone Awry: The Futile Quest for Artificial Intelligence," *Skeptic* 12, no. 2 (2006). http://www.skeptic.com/the_magazine/archives/vol12n02.html (August 20, 2008).
14. See Jeff Cares, *Distributed Networked Operations: The Foundations of Network Centric Warfare* (New York, NY: iUniverse, Inc., 2005).
15. Michael A. Palmer, *Command at Sea: Naval Command and Control since the Sixteenth Century* (Cambridge, MA: Harvard University Press, 2005), 302–03.
16. Headquarters, Department of the Army, *Counterinsurgency* (FM 3-24) (December 2006), 3–158.

Chapter 7. The Maritime Strategy of the 1980s

1. *The Evolution of the U.S. Navy's Maritime Strategy, 1977–1986* and *U.S. Naval Strategy in the 1970s: Selected Documents*, both edited by John Hattendorf and published by the Naval War College Press in 2004 and 2007 respectively. "U.S. Naval Strategy in the 1980s: Selected Documents" is in production, and selected documents from the 1950s and 1960s are also forthcoming.
2. James D. Watkins, *The Maritime Strategy* (Annapolis, MD: Naval Institute Press, 1986). This was the first open commercial publication, although Commo. Dudley Carlson had testified on the strategy in open session before the House Armed Services Committee three years prior.
3. Wesley McDonald, "The Critical Role of Sea Power in the Defense of Europe," *NATO's Sixteen Nations* 29, no. 2 (January–February 1985) 15.
4. Quoted in Steven E. Miller, "Rough Sailing: The U.S. Navy in the Nuclear Age," in *America's Defense,* ed. Michael Mandelbaum (New York, NY and London: Holmes and Meier, 1989), 198.
5. Quoted in Samuel P. Huntington, *The Common Defense: Strategic Programs in National Politics* (New York, NY: Columbia University Press, 1961), 369.
6. Scott MacDonald, "Inchon: The Impossible Landing," *Surface Warfare* (October 1981) 26.
7. Jeffrey G. Barlow, *The Revolt of the Admirals: The Fight for Naval Aviation, 1945–1950* (Washington, DC: Naval Historical Center, 1994).
8. Mahan, *Problem of Asia*, 26, 62–64.
9. Top Secret [Declassified], "Presentation To The President 14 January 1947 Vice Admiral Forrest Sherman, U.S. Navy, Deputy Chief Of Naval Operations (Operations)"; CNO Chronological File, Post 1 Jan. 46 Command File, Operational Archives, Naval Historical Center, Washington, DC. [emphasis in the original],
10. Thomas B. Hayward, "The Future of U.S. Sea Power," *Proceedings* 105, no. 5 (May 1979) 66–71.
11. Barry Posen, "Inadvertent Nuclear War," *International Security* 7 (Fall 1982), 28–54.
12. A two-part article entitled: "Back to the Stone Age," *Izvestia* (January 23–24, 1986), 5.
13. Martin Gilbert, *Winston S. Churchill*, vol. 3, *1914–1916* (London: Heinemann, 1971), 262.
14. See, in particular, Roger Thompson's *Lessons Not Learned: The U.S. Navy's Status Quo Culture* (Annapolis, MD: Naval Institute Press, 2007).

Chapter 8. Retrospective

1. Quoted in H. R. McMaster, *Dereliction of Duty* (New York, NY: Harper Collins, 1997), 72.
2. Alastair Iain Johnson, "Thinking About Strategic Culture," *International Security* 19, no. 4 (Spring 1995), 45.
3. http://thinkexist.com/quotation/no_man_ever_steps_in_the_same_river_twice-for_it/208269.html (August 22, 2008).
4. The distinction was made cogently by Lt. Col. Charles "Tony" Pfaff in his "Military Ethics in Complex Contingencies," in *The Future of the Army Profession,* ed. Don M. Snider, Lloyd J. Matthews, Jim Marshall, and Frederick M. Franks, 2d ed. (New York, NY: McGraw-Hill, 2005).
5. The issue is clearly framed, and the difficulties presaged, in Michael H. Hoffman, "Rescuing the Law of War: A Way Forward in an Era of Global Terrorism" *Parameters* (Summer 2005), 18–35. Of the Boumediene decision, Justice Antonin Scalia wrote, pungently, "It will almost certainly cause more Americans to be killed."
6. See: http://www.icgsdeepwater.com (August 18, 2007).
7. James T. Conway, Gary Roughead, and Thad W. Allen, *A Cooperative Strategy for 21st Century Sea Power*, (October 2007) http://www.navy.mil/maritime/MaritimeStrategy.pdf (August 22, 2008).
8. George C. Dyer, *The Amphibians Came to Conquer: The Story of Admiral Richmond Kelly Turner* (Washington, DC: U.S. Government Printing Office, 1972), 930.
9. Frank Hoffman, *The Marines: Premier Expeditionary Warriors* (Philadelphia, PA: Foreign Policy Research Institute, November 10, 2007) http://www.fpri.org/enotes/200711.hoffman.marinesexpeditionarywarriors.html (August 22, 2008).
10. Robert P Haffa and Robert E. Mullins, "Trends in America's Post-Cold War Military Conflicts: The Implications for Sea Power," *Sea Power* (July 2003), 14.
11. Charles G. Cooper, "The Day It Became the Longest War," *Proceedings* (May 1996), 79–80.
12. Lois E. Fielding, "Maritime Interception: Centerpiece of Economic Sanctions in the New World Order," *Louisiana Law Review* 53 (March 1993), 1191.
13. See Stephen J. Coughlin, "Modern-day Minehunting, Destroyer Style," *Proceedings* (June 2008), 42–46.
14. For more information on the Navy Expeditionary Combat Command see http://www.necc.navy.mil (August 22, 2008).

15. Quoted in Kaplan, "America's Elegant Decline," http://www.theatlantic.com/doc/200711/america-decline (August 22, 2008).
16. Typical of the genre is: Roger Cliff, Mark Burles, Michael S. Chase, Derek Eaton, and Kevin L. Pollpeter, *Entering the Dragon's Lair: Chinese Antiaccess Strategies and Their Implications for the United States* (Santa Monica, CA: The RAND Corporation, 2007).
17. Gray, *Recognizing and Understanding*, vii.
18. Colin S. Gray, "Strategy in the Nuclear Age: The United States, 1945–1991," in Williamson Murray, MacGregor Knox, and Alvin Bernstein, *The Making of Strategy: Rulers, States and War* (New York, NY: Cambridge University Press, 1994), 579.
19. John D. Hayes and John B. Hattendorf, eds., *The Writings of Stephen B. Luce* (Newport, RI: Naval War College Press, 1975), 131.
20. The "Pottery Barn Rule" in international law: "In the twentieth century, international law operated on the basis of you break it, you own it. If your country overthrows a government, then your country is responsible for creating and establishing a replacement government. Unfortunately, in the twenty-first century this rule is not working out very well (see the Balkans, Iraq, Palestine, Somalia and other cases). This idea needs to be rethought. The old model of conquering a country and then making it a vassal state against the will of its citizens is well past its expiration date. Some problems need to be eliminated by military force, but subsequent rebuilding may be best left to someone other than the original military organization." Garrett Jones, *Unconventional Approaches to Diplomatic Theory*, Foreign Policy Research Institute: E-Notes, http://www.fpri.org (March 30, 2007).
21. "When people see a strong horse and a weak horse, by nature they will like the strong horse." Osama bin Laden, "Transcript of Osama bin Laden Videotape, December 13, 2001, http://www.greatdreams.com/osama_tape.htm (August 22, 2008).
22. Hanson, *Carnage and Culture*, 416.
23. Alastair Finlan, *The Royal Navy in the Falklands Conflict and the Gulf War: Culture and Strategy* (London: Frank Cass, 2004), 13.
24. "Foreword to Jane's Fighting Ships, 1991–1992," quoted in *Seapower* (August 1992), 28.
25. Harvey C. Mansfield, *Manliness* (New Haven, CT: Yale University Press, 2006), 1.
26. Ibid., x.

27. Wayne Hughes Jr., "The Power in Doctrine," *Naval War College Review* 47, no. 3 (Summer 1995), 10.
28. Elaine Donnelly, "Constructing the Co-Ed Military" *Duke Journal of Gender Law and Policy* 14:815, 899.
29. Kingsley Browne, *Co-Ed Combat: The New Evidence That Women Shouldn't Fight the Nation's Wars* (New York, NY: Sentinel, 2007), 44.
30. Ibid., 176.
31. Lee Harris, *The Suicide of Reason: Radical Islam's Threat to the West* (New York, NY: Basic Books, 2007), 99.
32. Ibid., 101.
33. Ibid., 195.
34s. Robert D. Worley, *Shaping U.S. Military Forces: Revolution or Relevance After the Cold War* (Arlington, VA: Lulu, 2005), 197.

Chapter 9. Conclusion

1. In a letter to James Boswell, March 1759. Quoted in David A. Chappell, *Double Ghosts: Oceanian Voyages on EuroAmerican Ships* (Armonk, NY: M. E. Sharpe, 1997), 44.
2. Swartz, *Forward . . . From the Start.*
3. See: Peter M. Swartz, *U.S. Navy Capstone Strategies & Concepts (1970–2007): Insights for the Navy of 2008 & Beyond* (Alexandria, VA: Center for Naval Analyses, 2007).
4. Conway, Roughead, and Allen, *Cooperative Strategy*, 9.
5. J. C. Wylie, *Military Strategy: A General Theory of Power Control* (Annapolis, MD: Naval Institute Press, c. 1967), 162.
6. Thomas L. Friedman, *The World Is Flat: A Brief History of the Twenty-First Century* (New York, NY: Farrar, Straus and Giroux, 2006).
7. Victor Davis Hanson, "Do We Have a Strategy in the War? *National Review Online* (October 13, 2006).
8. Arthur Herman, *To Rule the Waves: How the British Navy Shaped the Modern World* (New York, NY: Harper Perennial, 2004), xvii.
9. Arleigh Burke, *ASW PAO Pamphlet* (Washington, DC: Department of the Navy, 1959).
10. George Modelski and William R. Thompson, *Seapower in Global Politics, 1494–1993* (Seattle, WA: University of Washington Press, 1988), 18.

11. Robert D. Kaplan, *Hog Pilots, Blue Water Grunts: The American Military in the Air, at Sea, and on the Ground* (New York, NY: Random House, 2007), 137.
12. Thomas Jefferson to John B. Colvin, September 20, 1810. *The Works of Thomas Jefferson*, collected and edited by Paul Leicester Ford. Federal Edition, 12 vols. (New York and London: G. P. Putnam's Sons, 1904–1905), http://press-pubs.uchicago.edu/founders/documents/a2_3s8.html (August 23, 2008).
13. Peter Karsten, *The Naval Aristocracy: The Golden Age of Annapolis and the Emergence of Modern American Navalism* (New York, NY: The Free Press, 1972), 53–54.
14. Ralph Peters, "Eyeing Iran," *New York Post* (January 6, 2007), 19.
15. London, *Victory in Tripoli*, 232.
16. Clark G. Reynolds, *History and the Sea: Essays on Maritime Strategies* (Columbia, SC: University of South Carolina Press, 1989), 8.
17. Thomas P. M. Barnett, "The Man between War and Peace," *Esquire*, April 23, 2008.
18. Bernard Brodie, *Sea Power in the Machine Age* (Princeton, NJ: Princeton University Press, 1941), 431.
19. George Santayana, *The Life of Reason: Or the Phases of Human Progress*, vol. 1 (New York, NY: Charles Scribner's Sons, 1906), 284–285.

Selected Bibliography

Alberts, David S., John J. Garstka, and Frederick P. Stein. *Network Centric Warfare: Developing and Leveraging Information Superiority*. Washington, DC: Center for Advanced Concepts and Technology, n.d.

Alden, Carroll Storrs and Ralph Earle. *Makers of Naval Tradition*. Boston, MA: Ginn and Company, 1925.

Alexander, Joseph H. and Merle L. Bartlett. *Sea Soldiers in the Cold War: Amphibious Warfare 1945–1991*. Annapolis, MD: Naval Institute Press, 1995.

Allen, Charles. *The Use of Navies in Peacetime*. Washington, DC: American Enterprise Institute, 1980.

Baer, George W. *One Hundred Years of Sea Power: The U.S. Navy, 1890–1990*. Stanford, CA: Stanford University Press, 1993.

Barash, David P. "Evolution, Males, and Violence." *The Chronicle of Higher Education*. May 24, 2002. http://chronicle.com/free/v48/i37/37b00701.htm.

Barlow, Jeffrey G. *The Revolt of the Admirals: The Fight for Naval Aviation, 1945–1950*. Washington, DC: Naval Historical Center, 1994.

Barnett, Roger W. "Naval Power for a New American Century." *Naval War College Review* (Winter 2002): 43-62.

____________. "Soviet Strategic Reserves and the Soviet Navy." In *The Soviet Union: What Lies Ahead?* edited by Kenneth M. Currie and Gregory Varhall. Studies in Communist Affairs, vol. 6. Washington, DC: U.S. Government Printing Office, 1984.

____________. "Strategic Culture and its Relationship to Naval Strategy." *Naval War College Review*, (Winter 2007): 24–33.

Bar-Yam, Yaneer. *Making Things Work: Solving Complex Problems in a Complex World.* n.P.: NECSI Knowledge Press, 2004.

Beach, Edward L. *The United States Navy: A 200-Year History.* Boston: Houghton Mifflin Co., 1986.

Beatty, Jack. "In Harm's Way." *The Atlantic*, May 1987, 37–53.

Beaufre, Andre. *An Introduction to Strategy with Particular Reference to Problems of Defense, Politics, Economics, and Diplomacy in the Nuclear Age.* New York, NY: Frederick A. Praeger, Publishers, 1965.

_____________. *Strategy of Action.* New York, NY: Frederick A. Praeger, 1967.

Beyerchen, Alan. "Clausewitz, Nonlinearity, and the Unpredictability of War." *International Security* 17, no. 3 (Winter 1992–1993): 59-90.

Binnendiyk, Hans, ed. *Transforming America's Military.* Washington, DC: National Defense University Press, 2002.

Bonds, John B. "Punishment, Discipline, and the Naval Profession." *Proceedings* (December 1978): 43–49.

Boslaugh, David L. *When Computers Went to Sea: The Digitization of the United States Navy.* Los Alamitos, CA: IEEE Computer Society, 1999.

Bracken, Paul. "Maritime Strategy and Grand Strategy," Zurich: ISN, September 28, 2006. http://www.isn.ethz.ch/news/sw/details_print.cfm?id=16726.

Breemer, Jan S. "The End of Naval Strategy: Revolutionary Change and the Future of American Naval Power." *Strategic Review*, (Spring 1994): 40–53.

Brodie, Bernard. *A Guide to Naval Strategy.* 5th ed. New York, NY: Praeger Books, 1965.

_____________. *A Layman's Guide to Naval Strategy.* Princeton, NJ: Princeton University Press, 1943.

_____________. *Sea Power in the Machine Age.* Princeton, NJ: Princeton University Press, 1941.

_____________. *Sea Power in the Machine Age.* New York, NY: Greenwood Press, 1943.

Brooks, Linton F. "Naval Power and National Security." In *The Use of Force,* edited by Robert J. Art and Kenneth N. Waltz. 3d ed. Lanham, MD: University Press of America, 1988.

Browne, Kingsley. *Co-Ed Combat: The New Evidence that Women Shouldn't Fight the Nation's Wars.* New York, NY: Sentinel, 2007.

Buell, Thomas B. *Master of Seapower: A Biography of Fleet Admiral Ernest J. King.* Boston, MA: Little, Brown, 1980.

Builder, Carl. *The Masks of War: American Military Styles in Strategy and Analysis.* Baltimore, MD: Johns Hopkins University Press, 1989.

Burk, James. "Military Culture." In *Encyclopedia of Violence, Peace, and Conflict,* edited by Lester Kurtz. Academic Press, 1999.

Burrell, Brian. *Damn the Torpedoes: Fighting Words, Rallying Cries, and the Hidden History of Warfare.* New York, NY: McGraw-Hill, 1999.

Calne, Donald B. *Within Reason: Rationality and Human Behavior.* New York, NY: Vintage Books, 1999.

Cares, Jeff. *Distributed Networked Operations: The Foundations of Network Centric Warfare.* New York, NY: iUniverse, Inc., 2005.

Cebrowski, Vice Adm. Arthur K. and John J. Garstka. "Network-Centric Warfare: Its Origin and Future." *Proceedings* (January 1998): 28–36.

Chernavin, Admiral of the Fleet V. "The Struggle for the Sea Lanes of Communication: Lessons of Wars and the Modern Era." *Morskoy Sbornik,* no. 2 (February 1990). Translated in JPRS-UMA-90-007, 23 March 1990.

Clary, James. *Superstitions of the Sea.* St. Clair, MI: Maritime History in Art, 1994.

Cliff, Roger, Mark Burles, Michael S. Chase, Derek Eaton, and Kevin L. Pollpeter. *Entering the Dragon's Lair: Chinese Antiaccess Strategies and Their Implications for the United States.* Santa Monica, CA: The RAND Corporation, 2007.

Cole, Bernard. *Oil for the Lamps of China.* McNair Paper. Washington, DC: National Defense University, 2003.

Conway, James T., Gary Roughead, and Thad W. Allen. *A Cooperative Strategy for 21st. Century Sea Power.* http://www.navy.mil/maritime/MaritimeStrategy.pdf, October 2007.

Coote, John O. ed. *The Norton Book of the Sea.* New York, NY: W. W. Norton & Company, 1989.

The Commander's Handbook On the Law of Naval Operations. NWP 1-14M. Washington, DC: Office of the Chief of Naval Operations, 2007.

Corbett, Julian S. *Some Principles of Maritime Strategy.* Annapolis, MD: Naval Institute Press, 1988 (originally published 1911).

Cotter, Charles H. *The Physical Geography of the Oceans.* New York, NY: American Elsevier Publishing Company, Inc., 1966.

Creveld, Martin van. *Technology and War: From 2000 B.C. to the Present.* New York, NY: The Free Press, 1989.

Cutler, Thomas J. *A Sailor's History of the U.S. Navy.* Annapolis, MD: Naval Institute Press, 2005.

Czerwinski, Tom. *Coping With the Bounds: Speculations on Nonlinearity in Military Affairs*. Institute for National Strategic Studies. Washington, DC: National Defense University, 1998.

Daveluy, Rene. *The Genius of Naval Warfare*. Annapolis, MD: Naval Institute Press, 1911.

Dismukes, Bradford, and James M. McConnell, eds. *Soviet Naval Diplomacy*. New York, NY: Pergamon, 1979.

Donnelly, Elaine. "Constructing the Co-Ed Military." *Duke Journal of Gender Law and Policy* 14:815, 2007.

Dorn, Edwin. *American Military Culture in the Twenty-First Century*. CSIS International Security Program report, access number 43311759.

Dyer, George C. *The Amphibians Came to Conquer: The Story of Admiral Richmond Kelly Turner*. Washington, DC: GPO, 1972.

Eliot, George Fielding. *Victory without War, 1958–1961*. Annapolis, MD: Naval Institute Press, 1958.

Elleman, Bruce A. and S. C. M. Paine, eds. *Naval Blockades and Seapower: Strategies and Counter-Strategies, 1805–2005*. London and New York, NY: Routledge, 2006.

Fenwick, Charles G. "The Freedom of the Seas." *The American Political Science Review* 11, no. 2 (April 1917): 386–388.

Finlan, Alastair. *The Royal Navy in the Falklands Conflict and the Gulf War: Culture and Strategy*. London: Frank Cass, 2004.

Firebaugh, Millard S., ed. *Naval Engineering and American Sea Power*. 2nd ed. Dubuque, IA: Kendall/Hunt Publishing Company, 2000.

Friedman, Norman. *Seapower and Space: From the Dawn of the Missile Age to Net-Centric Warfare*. Annapolis, MD: Naval Institute Press, 2000.

_____________. *U.S. Submarines Since 1945: An Illustrated Design History*. Annapolis, MD: Naval Institute Press, 1994.

Gaffney, H. H. *Globalization and U.S. Navy Forces*. CRM D0005743.A1/Final. Center for Strategic Studies. Alexandria, VA: Center for Naval Analyses, 2002.

George, James L., ed. *The U.S. Navy: The View from the Mid-1980s*. Boulder, CO: Westview Press, 1985.

Giamberardino, Oscar di. *The Art of War at Sea*. Washington, DC: Office of Naval Intelligence, 1958.

Gilje, Paul A. *Liberty on the Waterfront: American Maritime Culture in the Age of Revolution*. Philadelphia: University of Pennsylvania Press, 2004.

Glasser, Robert D. *Ballistic Missile-Carrying Submarines: A Reassessment of Their Contribution to Strategic Stability*. CISA Working Paper No. 68. Los Angeles,

CA: Center for International and Strategic Affairs, University of California, Los Angeles, 1989.

Goldrick, James, and John Hattendorf. *Mahan Is Not Enough: The Proceedings of a Conference on the Works of Sir Julian Corbett and Admiral Sir Herbert Richmond.* Newport, RI: Naval War College Press, 1993.

Gray, Colin S. *The Leverage of Sea Power: The Strategic Advantage of Navies in War.* New York, NY: The Free Press, 1992.

______________. *Modern Strategy.* Oxford: Oxford University Press, 1999.

______________. *The Navy in the Post-Cold War World: The Uses and Value of Strategic Sea Power.* University Park, PA: The Pennsylvania State University Press, 1994.

______________. *Recognizing and Understanding Revolutionary Change in Warfare: The Sovereignty of Context.* Carlisle, PA: U.S. Army War College Strategic Studies Institute, February 2006.

______________. *War, Peace, and Victory.* New York, NY: Simon and Schuster, 1990.

Gray, Colin S. and Roger W. Barnett. *Sea Power and Strategy.* Annapolis, MD: Naval Institute Press, 1989.

Gutman, Stephanie. *The Kinder, Gentler Military: How Political Correctness Affects Our Ability to Win Wars.* San Francisco: Encounter Books, 2000

Haffa, Robert P. Jr. and Robert E. Mullins. "Trends in America's Post-Cold War Military Conflicts: The Implications for Sea Power." *Sea Power* (July 2003): 13–16.

Hagan, Kenneth J. *This People's Navy: The Making of American Sea Power.* New York, NY: The Free Press, 1991.

Hanson, Victor Davis. *Carnage and Culture: Landmark Battles in the Rise of Western Power.* New York, NY: Doubleday, 2001.

Harris, Lee. *Civilization and Its Enemies: The Next Stage of History.* New York, NY: The Free Press, 2004.

______________. *The Suicide of Reason: Radical Islam's Threat to the West.* New York, NY: Basic Books, 2007.

Hattendorf, John B., ed. *Doing Naval History: Essays Toward Improvement.* Newport, RI: Naval War College Press, 1995.

______________, ed. *The Evolution of the U.S. Navy's Maritime Strategy, 1977–1986.* Newport Paper 19. Newport, RI: Naval War College Press, 2004.

______________, ed. *The Influence of History on Mahan.* Newport, RI: Naval War College Press, 1991.

_____________, ed. *Mahan on Naval Strategy: Selections from the Writings of Rear Admiral Alfred Thayer Mahan*. Annapolis, MD: Naval Institute Press, 1991.

_____________ and Richard Jordan, eds. *Maritime Strategy and the Balance of Power.* New York, NY: St. Martin's Press, 1989.

_____________, B. Mitchell Simpson III, and John R. Wadleigh. *Sailors and Scholars: The Centennial History of the U.S. Naval War College.* Newport, RI: Naval War College Press, 1984.

_____________. "The Uses of Maritime History in and for the Navy." *Naval War College Review* (Spring 2003): 12–38.

_____________. *U.S. Naval Strategy in the 1970s: Selected Documents.* Newport Paper 30. Newport, RI: Naval War College Press, 2007.

Hayes, John D. and John B. Hattendorf, eds. *The Writings of Stephen B. Luce.* Newport, RI: Naval War College Press, 1975.

Hegmann, Richard. "Reconsidering the Evolution of the U.S. Maritime Strategy 1955–1965. *Journal of Strategic Studies* 14, no. 3 (September 1991): 299–336.

Heinl, Robert Debs Jr.,. *Dictionary of Military and Naval Quotations.* Annapolis, MD: United States Naval Institute, 1966.

Herman, Arthur. *To Rule the Waves: How the British Navy Shaped the Modern World.* New York, NY: Harper Perennial, 2004.

Herrick, Robert Waring. *Soviet Naval Strategy: Fifty Years of Theory and Practice.* Annapolis, MD: United States Naval Institute, 1967.

Hoffman, Frank. *The Marines: Premier Expeditionary Warriors.* Philadelphia, PA: Foreign Policy Research Institute, November 10, 2007. http://www.fpri.org

Hoffman, Michael H. "Rescuing the Law of War: A Way Forward in an Era of Global Terrorism." *Parameters* (Summer 2005): 18–35.

Holland, John H. *Hidden Order: How Adaptation Builds Complexity.* Reading, MA: Perseus Books, 1995.

Holland, William J. Jr. "Where Will All the Admirals Go?" *Proceedings* (May 1999): 36–40.

Holmes, James. "Why Doesn't America Have a Nelson?" *Naval War College Review* (Autumn 2005): 15–24.

Holmes, Richard. *Acts of War: The Behavior of Men in Battle.* New York, NY: The Free Press, 1985.

Hornfischer, James D. *The Last Stand of the Tin Can Sailors.* New York, NY: Bantam Books, 2004.

Howarth, Stephen. *To Shining Sea: A History of the United States Navy, 1775–1991.* New York, NY: Random House, 1991.

Hughes, Captain Wayne P., Jr. "Naval Maneuver Warfare." *Naval War College Review* (Summer 1997): 25–49.

_____________. *Fleet Tactics: Theory and Practice.* Annapolis, MD: Naval Institute Press, 1986.

_____________. "The Power in Doctrine." *Naval War College Review* (Summer 1995): 9–31.

Huntington, Samuel P. *The Common Defense: Strategic Programs in National Politics.* New York, NY: Columbia University Press, 1961.

_____________. "National Policy and the Transoceanic Navy." *Proceedings* (May 1954): 483–93.

_____________. *The Soldier and the State: The Theory and Politics of Civil-Military Relations.* Cambridge, MA: Harvard University Press, 1957.

Isenberg, Michael T., *Shield of the Republic: The United States Navy in an Era of Cold War and Violent Peace.* Vol. 1, *1945–1962.* New York: St. Martin's Press, 1993.

Johnston, Alastair Iain. "Thinking About Strategic Culture." *International Security* 19, no. 4 (Spring 1995): 32–64.

Jones, Garrett. *Unconventional Approaches to Diplomatic Theory.* Foreign Policy Research Institute: E-Notes, www.fpri.org, March 30, 2007.

Kaplan, Robert D. "America's Elegant Decline." *The Atlantic Monthly* (November 2007):104–112.

_____________. *Hog Pilots, Blue Water Grunts: The American Military in the Air, at Sea, and on the Ground.* New York, NY: Random House, 2007.

Karsten, Peter. *The Naval Aristocracy: The Golden Age of Annapolis and the Emergence of Modern American Navalism.* New York, NY: The Free Press, 1972.

Keegan, John. *The Price of Admiralty: The Evolution of Naval Warfare.* New York, NY: Viking, 1989.

Klein, Yitzhak. "A Theory of Strategic Culture." *Comparative Strategy* 10 (1991): 3–23.

Kors, Alan Charles. "The West at the Dawn of the 21st Century: Triumph Without Self-Belief." *Watch on the West* 2, no. 1. http://www.fpri.org/ww/0201.200102.kors.westatdawn.html March 28, 2009.

Krepinevich, Andrew. *A New Navy for a New Era.* Report by the Center for Strategic and Budgetary Assessments, 1996. http://www.csbaonline.org/4Publications/PubLibrary/R.19960500.A_New_Navy_For_A_N/R.19960500.A_New_Navy_For_A_N.php

Laffin, John. *Women in Battle.* London: Abelard-Schuman, 1967.

Lambeth, Benjamin S. *American Carrier Air Power at the Dawn of a New Century.* Santa Monica, CA: RAND, 2005.

Lautenschlager, Karl. "Technology and the Evolution of Naval Warfare." *International Security* 8, no. 2 (Autumn 1983): 3–51.

Lehman, John F. *Command of the Seas.* New York, NY: Scribner, 1988.

Leonard, Henry A., Michael Polich, Jeffrey D. Peterson, Ronald E. Sortor, and S. Craig Moore. *Something Old, Something New: Army Leader Development in a Dynamic Environment.* MG-280. Santa Monica, CA: RAND, 2006.

Lewis, Kevin N. *Combined Operations in Modern Naval Warfare: Maritime Strategy and Interservice Cooperation.* P-6999. Washington, DC: The RAND Corporation, April 1984.

London, Joshua E. *Victory in Tripoli: How America's War with the Barbary Pirates Established the U.S. Navy and Built a Nation.* Hoboken, NJ: John Wiley & Sons, Inc., 2005.

Love, Robert B. Jr., ed. *The Chiefs of Naval Operations.* Annapolis, MD: Naval Institute Press, 1980.

Lynn-Jones, Sean M. "A Quiet Success for Arms Control: Preventing Incidents at Sea" *International Security* (Spring 1985): 154–184.

McCormick, Gordon H. and Mark E. Miller, "American Seapower at Risk: Nuclear Weapons in Soviet Naval Planning." *Orbis* 25, no. 2 (Summer 1981): 351–367.

McCrea, Michael M., Karen N. Domabyl, and Alexander F. Parker. *The Offensive Navy Since World War II: How Big and Why?* CRM 89-201. Alexandria, VA: Center for Naval Analyses, July 1989.

MacDonald, Scott. "Inchon: The Impossible Landing." *Surface Warfare* (October 1981): 22–31.

Mahan, Alfred Thayer. "Blockade In Relation To Naval Strategy." *Proceedings* (November 1895): 851–866.

____________. *The Influence of Sea Power upon History, 1660–1783.* London: Methuen and Company, 1965: first pub. 1890.

____________. *The Problem of Asia and Its Effect upon International Policies.* Boston, MA: Little, Brown, and Company, 1905.

Mahnken, Thomas. "Deny U.S. Access?" *Proceedings* (September 1998): 36–40.

Mansfield, Harvey C. *Manliness.* New Haven, CT: Yale University Press, 2006.

Maritime Strategy Implementation: The Conceptual / Intellectual Infrastructure. Newport, RI: Naval War College, 1987.

Melton, Buckner F., Jr. *Sea Cobra: Admiral Halsey's Task Force and the Great Pacific Typhoon.* Guilford, CT: The Lyons Press, 2007.

Merlin, Donald. *A Mind So Rare: The Evolution of Human Consciousness*. New York, NY: W. W. Norton & Company, 2001.

Mets, David. "Airpower and the Sea Services." *Airpower Journal* (Summer 1999): 67–85.

Miles, Edward L. Miles and Tullio Treves, eds. *The Law of the Sea: New Worlds, New Discoveries*. Honolulu, HI: The Law of the Sea Institute, William S. Richardson School of Law, 1993.

Miller, George H. *Provide for the Common Defense*. Washington, DC: [Washington Publications], 1988.

Miller, Steven E. "Rough Sailing: The U.S. Navy in the Nuclear Age." In *America's Defense,* edited by Michael Mandelbaum. New York, NY and London: Holmes and Meier, 1989, 194–230.

Mitchell, Brian. *Women in the Military: Flirting with Disaster*. Washington, DC: Regnery, 1998.

Modelski, George and William R. Thompson. *Seapower in Global Politics, 1494–1993*. Seattle, WA: University of Washington Press, 1988.

Moineville, Hubert. *Naval Warfare Today and Tomorrow*. Oxford, England: Basil Blackwell Publisher Limited, 1983.

Molnar, Thomas. *Authority and Its Enemies*. New Rochelle, NY: Arlington House Publishers, 1976.

Moriarty, Sean T. *Is the Concept of Sea-Based Logistics Sufficiently Specific to Enable Doctrine Development?* Newport, RI: Naval War College, 2002.

Morison, Samuel Eliot. "Notes on Writing Naval (not Navy) English." *The American Neptune* 9 (1949): 5–10.

Morville, Peter. *Ambient Findability*. Sebastopol, CA: O'Reilly, 2005.

Muir, Malcom. *Black Shoes and Blue Water: Surface Warfare in the United States Navy, 1945–1975*. Washington, DC: Naval Historical Center, 1996.

Murray, Williamson, MacGregor Knox, and Alvin Bernstein. *The Making of Strategy: Rulers, States and War*. New York, NY: Cambridge University Press, 1994.

Napoleon's Maxims. Translated by Col. S. C. Vestal and 1st Lt. F. J. Brunow of the Historical Section, Army War College, Washington, DC: May 1930.

Naval Studies Board, National Research Council. *Technology for the United States Navy and Marine Corps, 2000–2035: Becoming a 21st Century Force*. Washington, DC: National Academy of Sciences, 1997. http://www.nap.edu/html/tech_21st/tfnf.htm.

Neilson, Keith and Elizabeth Jane Errington. *Navies and Global Defense: Theories and Strategy*. Westport, CT: Praeger, 1995.

Nitze, Paul H., Leonard Sullivan Jr., and the Atlantic Council Working Group on Securing the Seas. *Securing the Seas: The Soviet Naval Challenge and Western Alliance Options*. Boulder, CO: Westview Press, 1979.

Noel, John V. *Naval Terms Dictionary*. 2nd ed. Annapolis, MD: United States Naval Institute, 1966.

O'Connell, D. P. *The Influence of Law on Sea Power*. Annapolis, MD: Naval Institute Press, 1975.

Owens, Bill. *Lifting the Fog of War*. New York, NY: Farrar, Straus, Giroux, 2000.

Owens, William A. "The Emerging U.S. System of Systems." *Strategic Forum* 63. Washington, DC: Institute for National Strategic Studies, National Defense University, February 1996.

Oxman, Bernard H. "The Territorial Temptation: A Siren Song At Sea," *The American Journal of International Law* 100:830, 841.

Palmer, Michael A. *Command at Sea: Naval Command and Control since the Sixteenth Century*. Cambridge Mass.: Harvard University Press, 2005.

_____________. *Undersea Warfare and Maritime Strategy: The American Experience*. Paper delivered at the Undersea Dimension of Maritime Strategy Conference, Halifax, Nova Scotia, June 22, 1989.

Parker, Kathleen. *Save the Males: Why Men Matter. Why Women Should Care.* New York, NY: Random House, 2008.

Parry, J. H. *The Discovery of the Sea*. Berkeley, CA: University of California Press, 1981; first pub. 1974.

Pinker, Steven. *Words and Rules: The Ingredients of Language*. New York, NY: Basic Books, 1999.

Ponce, Bryan P., *Hollow Promises: the Problem of Culture and the Integration of New Technology into the Navy*. School of Advanced Military Studies monograph. Ft. Leavenworth, KS: U.S. Army Command and General Staff College, 2004.

Posen, Barry. "Inadvertent Nuclear War." *International Security* 7 (Fall 1982): 28–54.

Potter, E. B., ed. *Sea Power: A Naval History*. 2nd ed. Annapolis, MD: Naval Institute Press, 1981.

Reason, J. Paul, with David G. Freymann. *Sailing New Seas*. Newport Paper 13. Newport, RI: Naval War College Press, 1998.

Reynolds, Clark G. *Command of the Sea: The History and Strategy of Maritime Empires*. 2 vols. Malabar, FL: Robert E. Krieger Publishing Company, 1983.

_____________. *History and the Sea: Essays on Maritime Strategies*. Columbia, SC: University of South Carolina Press, 1989.

Richardson, Elliot L. "Power, Mobility and the Law of the Sea." *Foreign Affairs*, (Spring 1980): 902–919.

Roach, Ashley. "The Law of Naval Warfare at the Turn of Two Centuries." *American Journal of International Law* (January 2000): 64–77.

Rodger, N. A. M., ed. *Naval Power in the Twentieth Century*. Annapolis, MD: Naval Institute Press, 1994.

Rose, Lisle A. *Power at Sea*. Vol. 3, *A Violent Peace 1946–2006*. Columbia, MO and London: University of Missouri Press, 2007.

Rosenberg, David Alan. "American Naval Strategy in the Era of the Third World War: An Inquiry into the Structure and Process of General War at Sea, 1945–90." In *Naval Power in the Twentieth Century*, edited by N. A. M. Rodger. Annapolis, MD: Naval Institute Press, 1994, 242–254.

____________. " 'It is Hardly Possible to Imagine Anything Worse' : Soviet Thoughts on the Maritime Strategy." *Naval War College Review* (Summer 1988): 69–105.

____________ and Floyd D. Kennedy Jr. *U.S. Aircraft Carriers in the Strategic Role*. Prepared for the Deputy Chief of Naval Operations (Plans and Policy) under Contract No. N00014-75-C-0237. Falls Church, VA: Lulejian & Associates, Inc., October 1975.

Rosinski, Herbert. *The Development of Naval Thought*. Edited with an Introduction by B. Mitchell Simpson III. Newport, RI: Naval War College Press, 1977.

Ryan, Paul B. *First Line of Defense: The U.S. Navy Since 1945*. Stanford, CA: Hoover Institution Press, 1981.

Schwarz, Benjamin C. *Casualties, Public Opinion, and U.S. Military Intervention: Implications of U.S. Regional Deterrence Strategies*. Santa Monica, CA: RAND, 1994.

Sea-Based Airborne Antisubmarine Warfare 1940-1977. Vol. I *1940–1960*. Prepared for Op-095 under ONR contract N00014-77-C-0338. Alexandria, VA: R. F. Cross Associates, 1978.

Sherwood, John Darrell. *Afterburner: Naval Aviators and the Vietnam War*. New York, NY: New York University Press, 2004.

Snider, Don M. "An Uninformed Debate on Military Culture." *Orbis* 43, no.1 (Winter 1999). http://search.ebscohost.com/login.aspx?direct=true&db=mth&AN=1497835&site=ehost-live (August 24, 2008).

Sommers, Christina Hoff. *The War Against Boys: How Misguided Feminism is Harming Our Young Men*. New York, NY: Simon and Schuster, 2000.

Sprout, Harold and Margaret. *The Rise of American Naval Power 1776–1918*. Annapolis, MD: Naval Institute Press, 1939.

_____________. *Toward a New Order of Sea Power: American Naval Policy and the World Scene, 1918–1922*. Princeton, NJ: Princeton University Press, 1946.

Sumida, Jon Tetsuro. *Inventing Grand Strategy and Teaching Command: The Classic Works of Alfred Thayer Mahan Reconsidered*. Washington, DC: The Woodrow Wilson Center Press and Baltimore: The Johns Hopkins University Press, 1997.

Swartz, Peter M. "Classic Roles and Future Challenges: The Navy after Next." In *Strategic Transformations and Naval Power in the 21st Century*, edited by Pelham G. Boyer and Robert S. Wood. Newport, RI: Naval War College Press, 1998, 273–305.

_____________. *"Forward . . . From the Start": The U.S. Navy & Homeland Defense: 1775–2003*. COP D0006678.A1/Final. Arlington, VA: Center for Naval Analyses, February 2003.

_____________. *Sea Changes: Transforming U.S. Navy Deployment Strategy: 1775–2002*. Arlington, VA: Center for Naval Analyses, 31 July 2002.

_____________. *U.S. Navy Capstone Strategies & Concepts (1970–2007): Insights for the Navy of 2008 and Beyond*. Briefing. Alexandria, VA: Center for Naval Analyses, 2007.

Symonds, Craig L. *Confederate Admiral: The Life and Wars of Franklin Buchanan*. Annapolis, MD: Naval Institute Press, 1999.

_____________. *The Naval Institute Historical Atlas of the U.S. Navy*. Annapolis, MD: Naval Institute Press, 2001.

_____________. *Navalists and Antinavalists: The Naval Policy Debate in the United States, 1785–1827*. Newark, DE: University of Delaware Press, 1980.

Tangredi, Sam J., ed. *Globalization and Maritime Power*. Washington, DC: Institute for National Strategic Studies, National Defense University, 2002. http://www.ndu.edu/inss/books/Books_2002/Globalization_and_Maritime_Power_Dec_02/32_ch31.htm (August 21, 2008).

Tazelaar, James, ed. *The Articulate Sailor*. Tuckahoe, NY: John de Graff Inc., 1973.

Terriff, Terry. "Innovate or Die." *Journal of Strategic Studies* 29, no. 3 (June 2006): 475–503.

Till, Geoffrey. *Maritime Strategy and the Nuclear Age*. New York, NY: St. Martin's Press, 1982.

Turnbull, Archibald D. and Clifford L. Lord. *History of United States Naval Aviation*. New Haven, CT: Yale University Press, 1949.

Uhlig, Frank, Jr. *How Navies Fight: The U.S. Navy and Its Allies*. Annapolis, MD: Naval Institute Press, 1994.

____________. *Navalists and Antinavalists: The Naval Policy Debate in the United States, 1785–1827.* Newark, DE: University of Delaware Press, 1980.

Vietnam: The Naval Story. Annapolis, MD: Naval Institute Press, 1986.

Walzer, Michael. *Just and Unjust Wars: A Moral Argument with Historical Illustrations.* n.p.: Basic Books, 1977.

Watkins, James D. *The Maritime Strategy.* Annapolis, MD: Naval Institute Press, January 1986.

Watts, Barry. *Clausewitzian Friction and Future War.* McNair Paper No. 52. Washington, DC: Institute for National Strategic Studies, 1999.

Weigley, Russell F. *The American Way of War: A History of United States Military Strategy and Policy.* Bloomington, IN: Indiana University Press, 1973.

Wildenberg, Thomas. *Dive Bombing, Midway, and the Evolution of Carrier Airpower.* Annapolis, MD: Naval Institute Press, 1998.

Williams, E. Cameron. "Sail Together or Sink Separately: The Convoy and Risk Analysis." *Defense Transportation Journal* (April 1987): 14–16.

Winnefeld, Adm. James A. and Dr. Dana J. Johnson. *Joint Air Operations: Pursuit of Unity in Command and Control, 1942–1991.* Annapolis, MD: Naval Institute Press, 1993.

____________. *Navalists and Antinavalists: The Naval Policy Debate in the United States, 1785–1827.* Newark, DE: University of Delaware Press, 1980.

"Why Sailors Are Different." *Proceedings* (May 1995): 65–70.

Woolsey, R. James. "Planning a Navy: The Risks of Conventional Wisdom." *International Security* (Summer 1978): 17–29.

Worley, D. Robert. *Shaping U.S. Military Forces: Revolution or Relevance after the Cold War.* Arlington, VA: Lulu, 2005.

Wylie, J. C. *Military Strategy: A General Theory of Power Control.* Annapolis, MD: Naval Institute Press, c. 1967.

____________. *Navalists and Antinavalists: The Naval Policy Debate in the United States, 1785–1827.* Newark, DE: University of Delaware Press, 1980.

"Why a Sailor Thinks Like a Sailor." *Proceedings* (August 1957): 811–817.

Yarbrough, Jean. "The Feminist Mistake: Sexual Equality and the Decline of the American Military." *Policy Review* (Summer 1985): 48–52.

Index

B

C

L

M

N

R

About the Author

Roger W. Barnett served in cruisers and destroyers and on oceangoing staffs during a twenty-three-year career in the Navy. His early career, in the Atlantic, took him on three deployments to the European Theater. More senior tours were served in the Pacific. He commanded the guided missile destroyer USS *Buchanan* (DDG-14) for twenty-five months, during which he participated in two deployments to the western Pacific.

Ashore, he served in four separate positions in the Pentagon, three of which were on the Navy (OPNAV) staff. While on active duty, he earned a PhD in international relations, writing his 1977 dissertation on Soviet military thought. On the OPNAV staff he was branch head in the Extended Planning Branch and of the Strategic Concepts Branch. The former prepared the Navy's Extended Planning Annex and the CNO's Program Assessment Memorandum; the latter, the Maritime Strategy. He was also deputy director of the Politico-Military Affairs Division. He retired from the Navy in 1984 as a captain.

He has published many articles in professional journals, the bulk of which have dealt with arms control and naval strategy. Along with Colin S. Gray he co-edited the book *Sea Power and Strategy*, published by the Naval Institute Press in 1989. His other book, *Asymmetrical Warfare,* was published by Brassey's in 2003. After serving on the faculty at the Naval War College, he retired as professor emeritus in 2001.